Facility Inspection Field Manual

Facility Inspection Field Manual

A Complete Condition Assessment Guide

Bernard T. Lewis

Richard P. Payant

McGraw-Hill

New York San Francisco Washington, D.C. Auckland Bogotá
Caracas Lisbon London Madrid Mexico City Milan
Montreal New Delhi San Juan Singapore
Sydney Tokyo Toronto

Library of Congress Cataloging-in-Publication Data

Lewis, Bernard T.
Facility inspection field manual : a complete condition assessment guide / Bernard T. Lewis, Richard P. Payant.
p. cm.
ISBN 0-07-135874-9
1. Engineering inspection—Handbooks, manuals, etc. 2. Plant maintenance—Handbooks, manuals, etc. I. Payant, Richard P. II. Title.

TA190.L42 2000
658.2′02—dc21

00-063838

McGraw-Hill

A Division of The ***McGraw-Hill*** *Companies*

1 2 3 4 5 6 7 8 9 0 DOC/DOC 0 6 5 4 3 2 1 0

ISBN 0-07-135874-9

The sponsoring editor for this book was Linda Ludewig and the production supervisor was Pamela A. Pelton. It was set in Century Schoolbook by Pro-Image Corporation.

Printed and bound by R. R. Donnelley & Sons Company.

This book is printed on recycled, acid-free paper containing a minimum of 50% recycled, de-inked fiber.

Contents

Foreword

The buildings in this country are aging much faster than they are being replaced. This means that these buildings and their components are expected to last longer, much longer, than their planned useful life. Consequently, many of these buildings are or will be pressed into services for which they were not designed. All of these factors, and others as well, will increasingly stress our facilities and their infrastructure and place even greater demands on maintenance professionals to deliver more value for each repair and maintenance dollar. TO MEET THIS CHALLENGE, UP-TO-DATE CONDITION STATUS OF THE BUILDINGS, EQUIPMENT, AND SYSTEMS MUST BE CONTINUALLY ASSESSED.

In this new and evolving arena, the best weapons any facility manager can have are current condition knowledge of the processes by which buildings and their equipment and systems interact, skillful use of the repair/replace model, knowledge of when and how to apply preventive and predictive maintenance, and most of all avoidance of inappropriate measures taken because the problem was misunderstood. It would be easy to predict catastrophe if the foregoing processes were allowed to reach their conclusion. However, catastrophe is not an acceptable outcome, so ways must be found to counter trends happening in the facility management environment.

Preface

Facility managers are stewards entrusted with maintaining and operating the organization's physical assets. As such, they must ensure that the funds they are allocated are efficiently and wisely used. The bottom line is that they need to get the "biggest bang for the buck" while supporting the overall mission of the organization.

For this reason, many facility managers today are asking questions such as: How do we preserve the organization's capital assets? How do we continue to operate and provide the required service? and How do we renew the current infrastructure? These are important questions that need to be addressed. In order for facility managers to support new and ongoing requirements they first need a starting point . . . which is to determine what assets they have and in what condition they are in.

A condition assessment (CA) is key to answering these questions and is necessary to enable informed decisions. This assessment is a process whereby the organization's facility systems, components, and subcomponents are evaluated as to their condition. The information obtained during the assessment is used to project costs for repair, renovation, or replacement. To simplify and standardize the process, i.e., to "match apples to apples," each facility must be evaluated using the same formal standards. Condition assessments are best done by architectural-engineering (A-E) firms. Most large firms are staffed to conduct these assessments but require a clearly defined scope of work; otherwise, expectations by the facility manager will not be met. Additionally, during this time of downsizing and austere in-house staffing, an in-house staff does not have the amount of time needed to conduct a thorough assessment. Higher management may realize this and, therefore, place more credence on an outside consultant's report. The in-house staff should be used to oversee and manage the A-E firm. Once the initial assessment is completed, it should be used as the starting point for a capital renewal program.

Once the CA is conducted and the capital program is established, an in-house inspection program can be used to update the plan annually. Planned, scheduled inspections will generate facility information which can be used to develop annual work plans and longer-range maintenance, repair, and replacement programs. Not only will this process identify capital maintenance requirements, but it will also result in the identification of routine maintenance deficiencies. It can also be used to predict impending equipment breakdown. An accurate and complete facility inspection is the foundation of any meaningful facility management maintenance program. Any type of inspection program is dependent on the experience and breadth of knowledge of the individuals conducting the inspection. Without accepted standards to follow, the potential is great that inspection results from various buildings, components, or systems or from inspections conducted from year to year will not be comparable. Not only will results not be comparable internally, they will not be able to be benchmarked.

The intent of this book, therefore, is to provide a logical, systematic approach to develop an in-house inspection program. It should be used as a practical tool in assessing and evaluating the condition of facilities and will result in a method of keeping an accurate inventory of assets. The checklists, which are the heart of this inspection program, will help in focusing on standards.

Bernard T. Lewis, P.E., C.P.E.
Richard P. Payant, CFM, C.P.E.

How to Use This Book

This book is meant to be used as a *personal* tool that summarizes, in a readily available manner, the sum total of all key, pertinent, and cognizant facility inspection assessment checkpoints needed by *facility managers, engineers, supervisors, property managers, home inspection technicians, and trade technicians* in the course of pursuing their day-to-day activities of managing and controlling the facility's operations and maintenance functional tasks. This book is meant to be used for either formal planned inspection tours or random supervisory work performance monitoring inspection.

The intent of this book is to be always at the facility manager, engineer, supervisor, or technician's *fingertips* (open at his or her desk, carried in a briefcase, or carried in a pocket). It should be used every day as an alternative to transporting bulky standards or technical specifications. It should be dog-eared, marked-up, and well used!

This book is designed with checklists that can, *in the field,* be quickly read, evaluated, and measured against predetermined operations and maintenance specifications criteria to determine issues at hand or potential long-term problems. Using this book will enable the quick detection and reporting of problems that require corrective actions. It simplifies the inspection assessment process by creating a pocket-sized reference book to be issued to all concerned so that *all are working on the same wavelength* in resolving maintenance and repair issues.

This book is not meant to be a *substitute* for professional expertise or detailed information found in other books of a more specialized and detailed nature. It is designed to serve as a continuing everyday readily available aid that presents the *summarization* of other available technical written resources.

Facility Inspection Field Manual

Part

1

Developing a Condition Assessment Program

Chapter

1

The Need For A Condition Assessment Program

Introduction

Facility managers today are plagued with requirements to save money. As such, benchmarking, downsizing, improving productivity and efficiency, streamlining supply, purchasing, and partnering are just a few of the initiatives ongoing in many facility departments today. “Doing more with less” is now a reality.

Traditionally, part of the annual facility funding request was based on reducing backlog in order to improve conditions within specific facilities. From a facility manager’s perspective, this was the best way to obtain funding; however, on the financial management side, it often fell on deaf ears. Facility managers, therefore, have had to become better versed in financial concepts in order to justify the financial benefit of the funding they sought. Today, they have to think in terms of what contributes to the organization’s bottom line. The more active role they can take in financial and budgeting matters, the more effectively their responsibilities can be accomplished.

Life of a Facility

Facilities are similar to the human body. They have a useful life, they age, live and breathe, and eventually they have final disposition. For example, the mechanical and electrical systems in a building are similar to the heart and lungs of the body; the energy management control systems are similar to the human brain; and the various environmental and life safety alarm systems are similar to the body's nervous system; the exterior envelope of the building corresponds to the body's skin.

As human beings take care of their bodies, they perform better; buildings are the same. If building systems are properly maintained and kept in good state of repair, their useful life is extended, they are actually cheaper to maintain, and less possibility exists for building sickness.

Life expectancy of a facility today is different than it was 20, 30 or even 50 years ago. Much of the difference is due to changes in types of materials used. A hundred years ago, slate was a common material used for roofs. A slate roof, although expensive to install, has a 100-year life expectancy. Today, however, other types of roof systems and materials are used, and these changes have changed the life expectancy of roofs. Mechanical systems were commercially introduced primarily after World War II (over 50 years ago). These systems have been redesigned and upgraded to the point, today, where in some cases it may be cheaper to dispose of the system than to try to repair or rebuild it.

Consequently, in this country we are seeing an aging infrastructure. "Deferred maintenance," in the last 10 years, has become a well-known concept, entailing enormous amounts of money to be expended. Good preventive and predictive maintenance programs and a method of inspecting facilities are important steps to slow down deferred maintenance growth.

As stated earlier, facilities are aging and are continuously being renovated, often in piecemeal fashion. These renovations can be for modernization and upgrade reasons or for change of mission. Many renovations are limited in scope due to funding restraints and many times result in cosmetic change with few or no infrastructure improvements. This fragmentation can lead to inefficiency of mechanical systems, customer complaints due to dissatisfaction with their facility or space environmental conditions, and eventually higher utility bills and maintenance costs. Consequently, recommissioning of the entire facility on a periodic basis (every few years) may be needed. This can be done by hiring a commissioning agent, but that will be expensive. As an alternative, it may be possible to free up an in-house mechanical engineer, who, using checklists in this book, could conduct periodic inspections of entire facility systems. This information can then be used to feed into either the organization's capital plan or the deferred maintenance list.

Knowledge Is Power

The knowledge and information possessed by the facility manager have the power to transform a reactive maintenance program into a proactive one. A proactive maintenance organization is one that optimizes future benefits, i.e., maximizes return on investment. Funding, therefore, has become the standard of performance—in other words, what value the organization is getting for its maintenance dollar. Because maintenance protects the capital investment of equipment, systems, and physical plant, it should be considered an investment in the future.

Inventory

To maximize return on investment of physical plant assets, the facility manager must have good information on the distinctive elements and status of the assets managed. A detailed inventory of the physical plant's equipment, and the components that make up that equipment, is imperative and forms the basis for preventive maintenance, predictive maintenance, space utilization, and capital asset replacement analysis.

Equipment, Systems, and Structures Inspections

Few facilities management activities have as much potential to influence facilities improvements and funding as formal scheduled inspections. Total building, equipment, and system inspections are an essential part of an overall facilities management plan. Facility managers should think of formal inspections not as a replacement for other types of maintenance and operations processes but as another process in the network of interdependent processes that make up the facilities management system. Data collection, deficiency reporting, and assessment activity for a piece of equipment is of very little value if the person performing the inspection does not understand how a specific piece of equipment fits into the system of which it is a part. Every piece of mechanical equipment must be part of an interdependent system that results in the temperatures, volumes, control, and automation that support the occupant's comfort and mission. The same can be said for electrical, life safety, and structural components. They are all parts of systems that are compromised when something is done to a piece without the knowledge of how all the pieces interrelate. Well-intentioned employees working without appropriate knowledge and understanding of the system can create many problems.

Planned Inspections

The condition of the facility and overall equipment and systems inventory must be periodically evaluated. This can be done through formal maintenance inspections. Formally planned and scheduled inspections can serve a variety of purposes:

- Confirming the effectiveness of preventive, predictive, and corrective maintenance programs
- Verifying that equipment and systems are operating as intended
- Collecting data
- Interacting with facility users
- Making formal equipment, systems, and structural deficiency assessments

Determining the goals of an inspection program well in advance of implementation allows for various types of planning and preparation.

While there may be many purposes, there are three primary categories of inspections that can assist facility managers in developing and monitoring facility management operations and determining facility conditions. An inspection can be conducted to determine the existence of system and equipment deficiencies and confirm that equipment and systems are functioning as intended. Inspections can also be organized around:

- Data collection tasks
- The need to communicate regularly with customers
- Being able to experience equipment, structures, and mechanical and administrative systems from the customers' perspective.

The challenges represented by the need to maximize the benefit of the time and expertise spent on inspections requires high-quality planning, training, and documentation. Personnel looking for deficiencies and checking on equipment and system settings require a high degree of technical training in the technology they are inspecting. Inspectors conducting data collection inspections, or collecting data in conjunction with other inspection activities, need to understand how the data will be used and how they relate to the functioning of the equipment and systems involved. Inspectors communicating with customers need to be trained in effective communication skills and need enough technical knowledge to explain equipment and systems to those who occupy the facility.

For real estate, the three most important words are *location, location, location*. For inspections to be successful, the three most important words are *planning, planning, planning*. Planning must precede the inspection and must also be continuous to ensure that the inspection methods are constantly improving and that the maximum amount of benefit is derived now and in the future.

Formal Inspections

Trained facility inspectors familiar with the facility and the equipment and systems, knowledgeable about maintenance standards, and having an overall interest in the facility can provide an evaluation of the components' remaining life expectancy. This evaluation is necessary for capital asset replacement planning and will also have a dramatic impact on the facility budget. These inspections are systematically planned and scheduled. Three types of formal inspections are normally included in an inspection-driven maintenance management program.

- Mechanic inspections
- Preventive and predictive maintenance inspections
- Facility inspections

Mechanic inspections

These are performed by a maintenance mechanic (sometimes by the equipment operator in manufacturing organizations), usually on a frequent basis, daily or weekly. Normally these are conducted at equipment startup, during operation, and at shutdown. These inspections include equipment cleaning, lubrication, and a visual check of belts, wiring, hoses, bushings, etc.

Preventive and predictive maintenance (PM&PdM) inspections

As equipment ages and deteriorates, the importance of preventive maintenance inspections is obvious: serious problems resulting from breakdowns can be reduced. Preventive maintenance inspections involve the verification that the PM was completed: lubrication, adjustments, and minor repair of equipment. Preventive maintenance guides, checklists, and schedules are needed in order to have an effective program. Predictive maintenance offers a means to make maintenance-related decisions on the necessity to take maintenance action, or no action, without reliability being sacrificed. It involves using available technology, appropriate to each piece of equipment, to set parameters to be measured against and ensure that resources are being used to their best advantage.

Facility inspections

These inspections are organized and planned visual examinations conducted by technically proficient personnel. The result of these inspections is a report depicting deficiencies for the various systems and their components and of the facility being inspected, priorities for maintenance, and the overall cost to be used for budgeting and planning purposes. Today, more companies are emphasizing technical expertise and knowledge of their personnel. Some organizations require employees to have certification, licensing, or professional registration. Numerous designation programs emphasize that success in a technical field, such as facilities management or engineering, requires special skills, knowledge, and values. Below is a partial list of organizations and the certifications they offer.

- Association of Facility Engineering (AFE)
  - Certified Plant Engineer (CPE)
  - Certified Plant Maintenance Manager (CPMM)
  - Certified Facilities Environmental Professional (CFEP)
- Building Owners Management Institute (BOMI)
  - Facility Management Administrator (FMA)
  - Real Property Administrator (RPA)
- International Facility Management Association (IFMA)
  - Certified Facility Manager (CFM)
- Institute of Real Estate management (IREM)
  - Certified Property Management (CPM)

Additionally, certifications and licenses are available for various trades which allow mechanics to demonstrate their expertise. Some certifications are required by law. For example, Section 608 of the Clean Air Act requires any mechanic or technician servicing or repairing a system having refrigerants such as chlorofluorocarbons (CFC, HCFC, or HFC) to be certified in accordance with the Environmental Protection Agency certification requirements. The National Institute for the Uniform Licensing of Power Engineers (NIULPE) is a national-level licensing agency that establishes uniform standards for firemen, water-tenders, power engineers, operators, examiners, and instructors. NIULPE has five levels of licenses: Chief Engineer, First Class Engineer, Second Class Engineer, Third Class Engineer, and Fourth Class Engineer. Some municipalities have their own licensing requirements, which include Journeyman- and Master-level licensing. Many have adopted the NIULPE standards. Finally, engineers are recognized in their profession when they receive their professional registration as a Professional Engineer (PE).

Sell the Maintenance Program

A facility inspection provides the facility manager with the hard data needed to sell the maintenance program. A consistent and objective inspection is powerful and is needed for budgeting. It enables managers to take the offensive to protect physical plant funding. It should contain enough detail to support required functions, identify the amount of work to be done, and show the total cost of labor, material, contracts, transportation, and all other aspects of how the funds will be expended.

A facility condition inspection generates an enormous amount of information and provides an overall snapshot of the current condition of the facility being examined. Facility inspections should be conducted on a routine basis, depending on personnel qualifications, funding, and time availability. All systems of an entire facility should be examined in detail. The facility inspection, then, establishes a baseline depicting

- The current level of maintenance accomplished to date
- The current operating condition of the system, equipment, or individual components
- The remaining useful life of the equipment and system

- Whether the equipment and system are operating within design parameters

Inspection Database

Normally, this information is then fed into a computerized maintenance management system (CMMS) (see Figure 1.1), where an information database can be developed and continuously updated and expanded. This process is accomplished in three phases:

1. The collection of information: specifically, the inspection itself, using checklists shown in Part 2 of this handbook, and annotation of specific observations.
2. Data entry: entering information from field notes into the database. This is a fairly straightforward undertaking that a clerk or inspector can accomplish.
3. Report generation. During this phase, various types of reports can be customized to meet requirements at hand.

Foundation of Inspection Program

The foundation of a facilities formal inspection program is a series of condition maintenance inspections. These inspections are performed by trained facility inspectors who identify deficiencies and can then estimate the cost of repairs. Essentially, they tell management the status of maintenance and the cost to fix. Another benefit is the continuous identification of deferred maintenance items. This helps to focus attention of financial decision makers on the magnitude of deferred maintenance and the importance of controlling it.

A one-time condition assessment, conducted by a consultant firm, is valuable to determine the total cost of deferred maintenance for the organization at that single point in time. But deferred maintenance is not static. Unless the database is continuously updated, it ceases to be an accurate depiction of the true amount of deferred maintenance. The formal inspection program and the checklists described in this handbook provide a simple method of keeping the database updated.

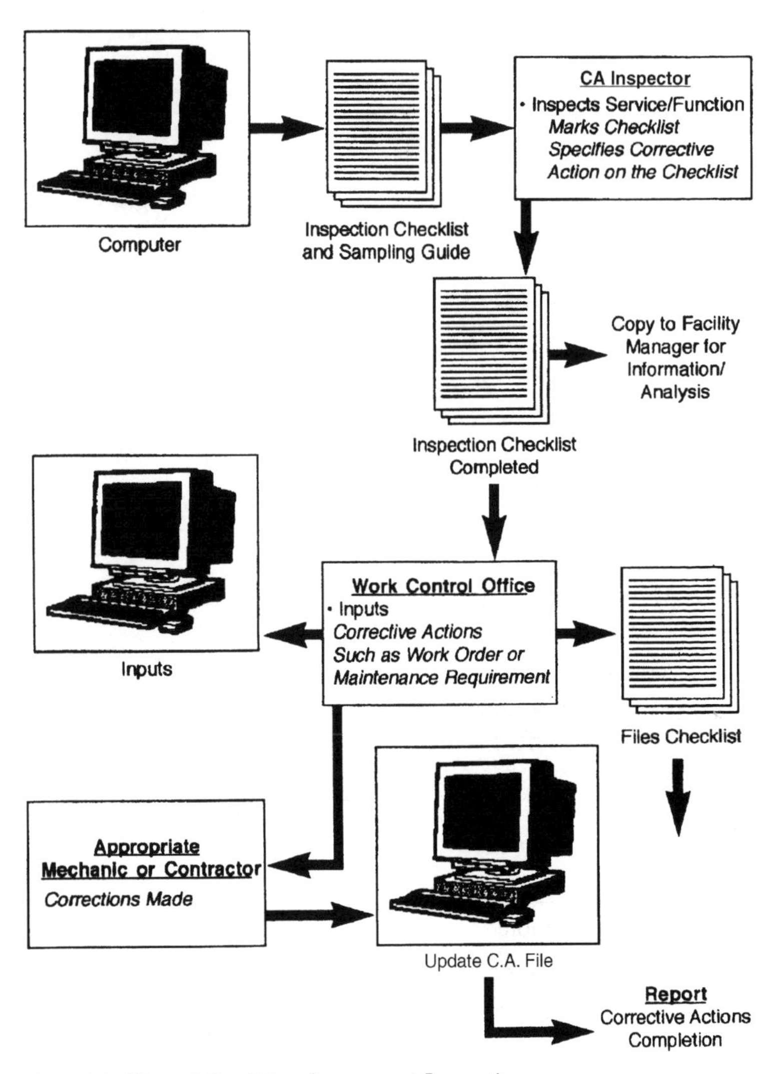

Figure 1.1 Flow of Condition Assessment Inspections

Chapter

2

The Condition Assessment Program

Developing a Condition Assessment Program

When a facility manager decides to develop a condition assessment program numerous questions and concerns need to be addressed up front. For example:

Question: Who performs the inspection?

Hire outside consultant or contractor, or use existing in-house staff? The information you want, as well as the cost, will drive the choice.

Question: Are members of your in-house staff qualified to inspect the various equipment, systems and their components, of the facility . . . and can they provide accurate, meaningful report data?

All inspection team members should be well trained in the inspection procedures and must be qualified to conduct the inspection. Licensing and certification criteria should be established and required.

Question: What is the detail level of the inspection?

Is it to examine mechanical and electrical systems, or is it to look at the "housekeeping/appearance" of the facility? This will drive the inspection interval selection.

Question: What is the planned interval between inspections?

The longer the interval, the more extensive the inspection should be. Annual facility inspections seem to be the most common for detailed system inspections.

Question: Are you inspecting a facility as part of a maintenance program?

Facilities must be kept in top condition in order to support the organization's mission. Inspections are one way to do this. Keep in mind that the longer the inspection interval, the more comprehensive the inspection must be.

Question: What portion of the facility will be inspected?

Access to various locations in the facility may be an issue, depending on the facility usage, time of day, time of year, etc.

Question: Is specialized equipment needed to access various locations or test system components?

Can you walk the roof of a shopping mall? Is a lift truck needed to examine walls and windows? Is rigging required for high-rise buildings?

Purpose of the Condition Assessment Program

- Gain a better understanding of the facility.
- Build the maintenance backlog and determine priorities.
- Identify impending deficiencies before they become major problems.
- Minimize system downtime.
- Extend the useful life of the facility
- Maximize energy efficiency.
- Help maintain property value.

- Identify long-term issues for capital planning and renewal (promote the deferred maintenance program and maintenance backlog).
- Assist in building transactions such as "due diligence."
- Provide better service to facility occupants.
- Improve communication among maintenance personnel, project managers, supervisors, engineers, etc.
- Enable better-trained maintenance workforce.
- Improve scheduling.

Implementation

The first step in implementing a successful inspection program is to decide exactly what the inspection is to accomplish. Inspections of mechanical rooms require equipment inventories and training to ensure that those conducting the inspection know and understand all of the equipment involved and understand the interdependencies of the systems they are inspecting. Equipment inventories usually incorporate some sort of equipment tagging system so that equipment can be identified easily by tag number. Tagging systems often are organized around codes for different types of equipment so that some knowledge about the type of equipment and possibly the location can be derived from the tag number. Even the simplest daily walk-through of mechanical spaces requires a broad range of understanding of the equipment and systems being inspected. An annual inspection of a building envelope requires knowledge about the construction of the facility and the dates of additions, remodeling, roof replacements, etc. Data collection inspections may require the installation of output monitors on systems and equipment and outcome monitors in occupied space. It is very important to determine what to measure when designing a data collection inspection. A significant amount of time can be spent collecting and analyzing data that cannot be developed into meaningful information. Measuring just the output of systems and pieces of equipment does not provide information concerning whether this output measure is having the desired result in the space the system and equipment serves. Inventory and assessment inspections require that inspectors know how to evaluate and identify equipment. If equipment will have to be taken out of service, downtime will have to be arranged with those served by the system.

Facility managers with computerized maintenance management systems (CMMS) will gain the most if they organize the instructions and scheduling of their inspections within the same system used to organize other types of facilities work. These systems can be configured to provide specific instructions, including tools and equipment lists to ensure that inspectors have what they need when they get on-site. Nothing is more expensive than multiple mobilizations to perform one job.

Those without a CMMS will have to prepare work assignment sheets consistent with whatever system they are using for the organization of other work. It is important to integrate inspections into the routine of the facility management organization. These inspections have to be important to management for them to be important to others. If the work is not well organized and not integrated into systems used for other "important" work, the message is loud and clear.

Any task assignment sheets need to be organized so that information about the tasks to be performed is clearly laid out and logical. A building inspection involving several mechanical rooms should be organized so that inspectors can work down a single sheet or through several sheets as they go through the inspection and not be required to flip around a page or back and forth between pages. The sheets should also include enough space for inspectors to record their findings. Boxes provided for the recording of gauge readings should be large enough to accept likely readings. Whenever possible, instructions should help the inspector identify readings or conditions that are abnormal and indicate how to respond to those conditions. Instructions may include a range of readings that is considered normal for a particular gauge. When an abnormal condition is experienced, the inspector must know how to respond. Equipment may need to be de-energized and locked out, and emergency calls may need to be made. Equipping inspectors with the right emergency information is extremely important.

Inspections designed as a purely assessment activity require significant research into the equipment and systems under review. For these inspections, installation date and maintenance history are critical to successful assessments. Methods of expressing condition assessments also need to be determined prior to conducting this type of inspection. Usually the purpose of assessment inspections is to determine what work is required, including replacement, and when work is to be done. In the normal world of facility management budgets, there is usually a need to establish replacement and major repair priorities, and valuable information for this process is available during an assessment inspection. Determining that a piece of equipment needs to be replaced is usually not enough. Even determining that it needs to be replaced within the next six months is not enough. A useful way to express more information about condition of a piece of equipment is to express it as "percent of useful life left." This piece of information may assist in the difficult job of comparing and establishing priorities among many dissimilar pieces of equipment and/or building components. Inspectors making judgments like these need to be supported with information and training. If information is being gathered through the work of a number of inspectors, the reliability of the assessments is a function of the quality of the inspector training. This is surely one type of training that will benefit from a professional needs analysis and presentation rather than being left to the last mechanic or engineer to go through it. The validity of these assessments can also be called into question if the judgement is based solely on the experience or expertise of the inspectors. Accuracy in these judgments is greatly facilitated by the use of manufacturer's operations and maintenance manuals and guidelines about useful life. If you know that a piece of equipment has been in place for five years, has benefited from a good preventive and predictive maintenance program, has had few maintenance problems, and is expected to last 10 years, then you can reasonably expect the equipment to last another 5 years or have

50% of its useful life left. It also needs to be mentioned that effective preventive and predictive maintenance can often extend the life of a system or piece of equipment well beyond that predicted by the manufacturer. Equipment that is monitored on a regular basis can provide data that may indicate that the frequency of preventive and predictive maintenance and/or data collection inspections should be altered. This is especially true if equipment is being kept beyond its recommended life.

An often-overlooked step in the implementation of an inspection program is to ensure that the facility inspector has access to all the areas to be inspected. It is necessary to equip inspectors with whatever keys, codes, and/or access cards they will need to access the space required.

It is also necessary to establish logical inspection routes so that the inspector can move logically from one area to another without needing to double back and travel past facilities that will be inspected later. When inspections need to be conducted in adjacent facilities, they should be scheduled at the same time. Establishing inspection routes is often a good strategy so that whole areas of a facility can be inspected at the same time. This factor is especially important with complex facilities like a university campus or municipality where the distances between facilities can be significant.

The topic of inspector training is important to view not only as an implementation step, but as part of a continuous improvement strategy to ensure that inspectors always know what they need to know. Training should be a systematic process that is well integrated into the overall facilities management plan. Inspector training for facility inspectors involves the same steps as training in other arenas. Briefly, it is composed of the following steps:

- Job analysis
  - Establish all required competencies.
- Needs assessment
  - What competencies do trainees have and what do they need training for?

- Basic training competencies
  - Given the variety in trainee competencies, what topics will be covered in training?
- Scheduling and training
  - When and where, type of facilities, length?
- Conducting training
  - Logistics, trainer readiness?
- Evaluation of training effectiveness
  - Evaluation of individual trainee?
  - Evaluation of trainee groups?
- Evaluation over time/trainers/schedule/etc.

Scheduling of Inspections

While the scheduling of inspections may seem a simple process, when done poorly it can have a dramatic impact on results. There must be different scheduling strategies for the different types of inspections. Daily walk throughs of mechanical areas are best done at the beginning of the day but benefit greatly from following a previous shift debriefing session. An inspection to determine if HVAC equipment is meeting the needs of the building occupants is best made during the time when occupants are using the facility. Inspections that require HVAC equipment to be taken off-line are best done when the occupants are not using the facility. Effective inspection scheduling is dependent on clear understanding of what the inspection is intended to accomplish. Inspections that involve the use of testing equipment or photography must be scheduled when this equipment can be used to best advantage. A thermal screening of a building exterior may best be done at night so that thermal readings of a facility are not affected by ambient heat energy. A thermal screening of an electrical panel may best be done during the day or at the time when the usage of electricity through subject panel is high. Scheduling must be carefully thought through and viewed as an integral part of the planning and operational process.

Staffing for Inspections

Two fundamental approaches to staffing are possible when planning an inspection program:

1. Determine the overall need of your facilities, design a system to meet those needs, and then determine where the facility inspectors will be obtained.
2. Determine what resources you have available and design a system to take advantage of what is available.

If you can tolerate the emotional pressure of a system that generates work you know you cannot complete, the former approach to staffing generates a lot of data in support of requests for additional human resources. If you have to use only what you have, you must design a system that focuses on the most important equipment and systems. You are the only one who knows what is the most important for your facility, but cost of equipment, cost of downtime, life safety, and criticality of mission are a few items that have to be given serious consideration. As a minimum, and depending on the type and size of facility, you should have inspection representatives that can examine the building's structural components and mechanical and electrical systems. Preferably, these facility inspectors should be qualified, experienced engineers, but you can also include your trade foremen in order to obtain the true, actual status of the facility. The mix and qualifications of facility inspectors should be matched with their expertise.

Who Should Inspect

Using consultants or contractors to conduct periodic condition assessments ultimately costs "big bucks" and is only accomplished once a year or every several years. Why not establish a self-sustaining maintenance management program? Who better knows the condition of your facilities than your own workforce? Who, other than your workforce, has a major stake in the operation and maintenance of your facilities? Does a consultant or contractor that you hire one time know all the idiosyncracies of your facilities?

Where will the consultant go for much of the information that is needed? Remember: consultants are at your facility to complete a job, turn over a document of some type, and depart with a profit. In-house inspections have two advantages: they save money and ensure confidence. On the other hand, hiring outside expertise reduces in-house involvement and allows in-house personnel more time to perform their jobs. Consultants who specialize in specific trade areas are generally well qualified and because of their vast experience and familiarity with specific systems can hone their diagnosis to potential trouble spots.

Conducting condition assessments with in-house staff can save money, but personnel require the knowledge of the equipment and systems being inspected and the interest of doing the inspection. Firsthand knowledge of the equipment and systems allows the facility manager and inspectors to have meaningful discussions concerning repairs. When technology is introduced, the detail of the inspections expands. The cost of inspections is cheap when balanced against the data gathered. Inspections are a critical step toward enhancing the life of a facility.

Some types of inspections and surveys, because of their technical nature, lend themselves to the use of outside consultants. Some examples:

- Thermographic roof inspections
- Ultrasonic underground storage tank inspections
- Thermographic electrical distribution panel inspections
- Eddy current testing of chiller tube bundles
- Elevator full-load tests
- Ergonomic surveys
- Lighting power factor studies
- High-voltage main switchgear inspections
- Steam system surveys
- Chiller overall assessments.

Inspector Qualifications

Facility inspectors shall be either engineers with degrees, as appropriate, in the civil, electrical, mechanical, or architectural disciplines, or journeyman craftsmen, as appropriate, in the structural, electrical, mechanical, or heating, ventilating, and air conditioning disciplines. Both the engineers and the craftsmen should have experience and working knowledge of two other disciplines.

Inspector Responsibilities

- Conduct thorough examinations using sufficient time to complete a condition inspection checklist with accurate and reliable data. When a thorough examination is not possible, this fact should be reported.
- Reach conclusions regarding facilities condition based on the scoring instructions and on personal observations and analysis.
- Make every reasonable effort to determine the true cause of a deficiency.
- Review for acceptable quality levels of previously made repairs.

Condition Assessment Information Sources

Equipment and system inventories, and general facility information, can initially be obtained from several sources:

- Manufacturer's recommendations, which are usually annotated in operational and all internal equipment manuals
- Building construction or renovation specifications, which are part of the construction contract
- "As-built" drawings, which are the final project plans, annotated with changes made in the field
- Governmental specifications available from federal, state, county, and/or city jurisdictions

- Preventive and predictive maintenance database
- Customer input
- Warranty information
- Deferred maintenance lists
- Known deficiencies (ADA, life safety, etc.)
- Facility coordinators
- Maintenance and custodial workforce
- Outstanding issues from recommendations of past inspections
- Occupational Safety and Health Act (OSHA)

Planning the Inspection

Once the timeframe for the inspection has been identified, coordination and notification must occur between facilities personnel and occupants/customers. Below are outlined some of the items that must be considered.

- *Scheduling the inspection.* Various groups and organizations should be queried concerning any special events that might be impacted or laboratory experiments that might be ongoing and cannot be postponed.
- *Notification.* Occupants/customers should be given advance notice of the inspection. This can be done using various methods, such as voice mail, e-mail, telephone, and flyers posted at different locations throughout the facility giving a point of contact to call for any questions that may arise.
- *Training of inspectors.*
  - Review of information (tools)
  - Review of standards to follow
  - Responsibilities of each inspector
  - Recording deficiencies (keeping field notes and use of camera/video camera)
  - Estimating ballpark costs

- *Definition of clear, succinct deficiencies.*
- *Equipment needed during the inspection.*
  - Measuring tape
  - Notebook
  - Flashlight
  - Tape recorder
  - Camera
  - Hand tools
  - Reduced facility plans
  - Key access
  - Rags
  - Results of previous inspections

Inspection Safety

Appropriate safety equipment, such as safety glasses, hard hats, gloves, respirators, and safety belts, shall be provided to each inspector in accordance with the nature of the hazards to be confronted. The Safety Director (or equivalent) shall be consulted concerning the protective equipment required, the maintenance and use of such equipment, and safety precautions to be observed during the normal course of inspections. Inspectors shall carefully observe all safety precautions in the conduct of inspections and tests to avoid hazards to themselves, others, and provide an example to other personnel working on or near the item being inspected. OSHA standards, as a minimum, are to be adopted as safety criteria for maintenance deficiency reporting activities. All inspectors shall become well versed in all pertinent safety instructions.

Inspector Assignments

The following table outlines primary responsibilities for performance of the facility condition inspection, by discipline.

Assignments		
Assessment Area	Discipline	Craftsman
Structural	Civil engineer	Carpenter
Electrical	Electrical engineer	Electrician
Mechanical	Mechanical engineer	Plumber, pipefitter, operating engineers
Heating, ventilating, and air conditioning	Mechanical engineer	HVAC mechanic, operating engineers

Documentation

Proper documentation is a critical aspect of any inspection program. Excellent computer programs do exist to assist with customizing inspection reports; however, they are not absolutely necessary. What is important is that a database of this information should be recorded, measured, and continuously reported. As a minimum, reports should include:

- Scaled set of plans of the facility
- Facility equipment inventory
- Summary of checklist issues/comments/observations: equipment condition and equipment summary
- General summary of the inspection
- Composition of the inspection team
- Date/time of inspection
- Any code deficiencies
- Anticipated service life remaining
- Cost estimates
- Weighted scoring sheet
- Recommendations

Equipment and System Logs

Collecting data on the output and outcomes achieved by equipment and systems and on the resources consumed in the collection process is the first step in documenting not only the performance of systems, but the operation of the entire inspection program. Data collected and stuffed in a drawer or entered into a computer and forgotten are of no value and hardly worth the time it took to collect. Equipment and system logs provide the tools for technical, administrative, and management personnel to evaluate equipment, systems, and programs. Logs that lend themselves to statistical analysis are extremely valuable. Statistical analysis, such as statistical process control, can be used to establish ranges of normal equipment operation. These can guide preventive and predictive maintenance frequencies and analyses of operating parameter trends that could indicate problems before they affect a facility's occupants. Often, techniques such as these can identify problems whose detection prior to the implementation of a system inspection was available only after a failure of the system causing a service interruption to facility occupants. Noncomputerized logs, while less valuable than computerized logs, can play an important role because some of the information that can be derived from log data can be seen at a glance once all of the data are logged in one place. Log sheets kept in equipment rooms are often used as quick references to trends in operating parameters over time. Noncomputerized equipment and system logs as permanent records must be maintained in locations where they can be preserved for the life of the equipment or system. Decisions that need to be made about replacement of equipment and systems are often well informed by a review of the logs kept on the equipment or system being replaced.

Technology

The advent of energy management systems and CMMS has resulted in a focus on the value of formal inspections. While one can make a "virtual" inspection of a facility through an energy management system's screens, it does not replace the trained eye, ears, and touch of a trained and experienced inspector. An energy management system is also a system that requires preventive maintenance and monitoring to keep it operating at its best. Energy management systems have become integrated into routine facility management activity, and inspections can add value to the technological system. Many CMMS's on the market have the capability to organize all types of data collected through the use of bar codes. This technology has had a dramatic impact on the quality and quantity of data that can be collected and the ease with which these data can be integrated in decision support systems. Most of us experience bar code technology each time we go to the supermarket. This point-of-sale data-gathering technology allows many retail businesses to keep daily track of inventory and analyze trends in product sales, while decreasing the amount of time it takes to go through the checkout and improving the accuracy of the data being entered. Inspections can be made more productive through the use of bar code technology. Imagine the value of having every piece of equipment tagged with a bar code so that identifying the equipment is accomplished by a mere swipe of a pen-sized device over the bar code sticker attached to the equipment. Any routine activity, scaled gauge readings, times, and temperatures can have a bar code identity to be swiped as needed to collect the data. Upon the completion of an inspection, the data can be downloaded into a computer system and the analysis can begin automatically. Data collected at night can be ready for analysis first thing in the morning of the next day. Another type of technology that can add value to an inspection system is scanning equipment. Scanners can be set up to scan field data collection forms and inspection reports directly into a computer system. Careful training must be done to help inspectors understand what the scanner can and cannot read, but this can be powerful technology when applied well.

Condition Assessment Procedures

Inspectors should follow a prescribed standard when examining equipment, systems, and components of those equipment systems. The average useful life of facility components is shown in Appendix 1. If defects are found, the severity of the defects should be determined. Each inspection should begin with the interior of the facility, then proceed to the building envelope. If conducted by in-house personnel, the following is a recommended procedure that has worked for several organizations.

- Once a week, a walk-through inspection of one or more facilities is conducted to review the material condition of the structure. The inspection party can consist of department engineers, trade managers, and occasionally maintenance workers. Inspections will be scheduled so that each facility is examined annually.
- One individual in the facilities department be responsible for coordinating and scheduling inspections and contacting the inspection team on the day of the inspection. This individual should also prepare a report of the discrepancies noted, cite responsibility for corrective action, and ensure that work is scheduled and tracked to completion. For convenience, inspectors can use hand-held portable tape recorders to record inspection results. More recently, hand-held computers have been available to be used by inspectors to complete inspection sheets and annotate issues. The information can then be downloaded and added to the inspection database. Work generated as a result of these inspections can then be used in developing the facility department's annual work plan.
- The facility condition inspection report will also be used to update the deferred maintenance backlog list, where appropriate. It will also be used as the tool to feed information into the capital program.

The key to any good inspection program is knowing what to look for and being able to measure and benchmark the results in order to determine the current state of your facilities. The checklists shown in Parts 2 and 3 of this book will be of great benefit. They can be used by in-house inspectors and by outside consultants/contractors as well. The checklists in Part 2 are organized in the following Facility Component Descriptions.

I. Building Systems Assessment
II. Operational Facilities Assessment
III. Utilities and Ground Improvement Assessment
IV. Utilities Plants
V. Fire Protection System Identification
VI. Special Systems Assessment

The checklists in Part 3 are abbreviated and less detailed. They can be used to conduct a "quick and dirty" inspection in order to obtain a brief synopsis evaluation of your facilities. Supplementary assessment inspection checklists covering topics involving safety, custodial, grounds, environmental, indoor air quality, predictive maintenance and energy management can be found in Part 4.

Checklist Scoring

Part 2

Scoring of these inspection checklists is simple. In Part 2 the inspector decides whether the checklist item is *emergency, urgent, routine,* or *deferred.* Simply place a checkmark in the appropriate column. At the end of the section being inspected, total the checkmarks in each column and divide each column total by the total number of items in that checklist. The result is a percentage of *emergency, urgent, routine,* or *deferred* items. Once the inspection is completed, a "checklist summary" is compiled, as shown in Figure 2.1. Here the totals of all component checklist items used are added and divided into the total column count for *emergency, urgent, routine,* and *deferred.*

Part 3

In Part 3 the inspector places a checkmark in the column marked "G" for *good,* "F" for *fair,* and "P" for *poor* opposite the

INSPECTION CHECKLIST SUMMARY					
COMPONENTS	TOTAL	EMERGENCY	URGENT	ROUTINE	DEFERRED
I. Building Systems Assessment					
Chimneys & Stacks	15				
Buildings/Structures	47				
Roofs	17				
Trusses	11				
Trailers	8				
Air Conditioning Syst.	106				
Cranes and Hoists	43				
Elevators/Platform Lifts/ Dumbwaiters	74				
Food Preparation and Service Equipment	20				
Heating Systems	4				
Plumbing Systems	22				
Ventilating and Exhaust Air Systems	14				
Hot Water Systems	10				
Electrical Systems	18				
Lighting	20				
Switch Gear	33				
I. SUB TOTAL	462				
INSPECTION CHECKLIST SUMMARY					

Figure 2.1 Inspection Checklist Summary

checklist item. At the end of each section the number of checkmarks in each column (*good, fair, poor*) is added and divided by the total number of checklist items to obtain a percentage of that condition.

COMPONENTS	TOTAL	EMERGENCY	URGENT	ROUTINE	DEFERRED
II. Operational Facilities Assessment					
Towers, Masts, and Antennas	17				
Chemical / Fuel Facilities	20				
Chemical (Fuel Facility Storage)	22				
Brows and Gangways	10				
Camels & Separators	9				
Dolphins	12				
Piers,Wharves,Quay-walls, & Bulkheads	30				
Disconnecting Switches	21				
Electrical Grounds & Grounding Systems	12				
Electrical Instruments	13				
Electrical Potheads	7				
Electrical Relays	5				
Lightning Arresters	19				
Power Transformers, Deenergized	25				
Power Transformers, Energized	36				
Safety Fencing	20				
Steel Poles/Structure	14				
Vaults and Electrical Manholes	21				
Cathodic Protection Systems	13				
Electrical Motors and Generators	18				
Pier Circuits and Receptacles	19				
Distribution Transform-ers, Deenergized	14				
Distribution Transform-ers, Energized	18				
Buried & Underground Telephone Cable	11				

Figure 2.1 *(Continued)*

Telephone Substation	21				
Fuses & Small circuit breakers(600V and below, 30A and below	8				
Rectifiers	10				
II. SUB TOTAL	445				
III. Utilities and Ground Improvements Assessments					
Bridges and Trestles	26				
Storm Drainage Syst.	14				
Fences and Walls	15				
Grounds	17				
Railroad Trackage	18				
Pavements	19				
Retaining Walls	8				
Tunnels and Underground Stucture	15				
Piping Systems	12				
Steam Distrib. Equip.	18				
Pumps	31				
Fresh Water Supply & Distrib. System	23				
Sewage Collection & Disposal Systems	19				
Unfired Pressure Vessels	33				
Underground Tanks	16				
III. SUB TOTAL	284				

Figure 2.1 *(Continued)*

Summary

It is of utmost importance to follow through on the assessment of your facilities and present the results to the responsible financial manager or decision maker. Using the results of the assessment, coupled with a strong executive summary and pertinent backup material, facility managers are in a better position to compete for funds with all the other financial demands placed on an organization.

INSPECTION CHECKLIST SUMMARY					
COMPONENTS	TOTAL	EMERGENCY	URGENT	ROUTINE	DEFERRED
IV.Utilities Plants					
Freshwater Supply	12				
Electrical Generation and/or Distribution	53				
Heating Generation	103				
Air Conditioning Generation & Distribution	166				
IV. SUB TOTAL	334				
V. Fire Protection Systems					
Fire Alarm Panels	3				
Building Sprinkler Alarm Valves	14				
Fire Alarm Boxes	30				
Security & Intrusion Alarm Systems	2				
V. SUB TOTAL	49				
VI. Special System Assessment					
Structural Features	2				
Electrical Systems	9				
Plumbing/Piping Syst	2				
Motor Assemblies Electric Motors	1				
Fans, Fan shafts, & FanShaft Bearings	1				
VI. SUB TOTAL	15				
TOTAL:	1589				

Figure 2.1 *(Continued)*

Ways to Make Your Point

When presenting your assessment information and attempting to make your point for financial support to improve facilities, make use of props. The old saying "A picture is worth a thousand words" will go a long way to support your case. For example: just presenting inspection results and costs to replace an air handler may not be enough. A more striking method, which will have a dramatic impact, is to show a piece of the paper-thin chill water coil and explain how it is only a matter of time until a catastrophe occurs which will have an impact on the organization's mission. If nothing else, make use of photographs or video clips to show a cross section of problem areas and why funding is needed. Also, as part of your presentation, it is advisable not only to show the problem areas but to depict how funding for past projects has benefited the organization. Some additional reasons to justify funding problems identified from facility condition assessments include:

- Less breakdown maintenance, resulting in fewer emergencies
- Improved maintenance worker morale, because they see corrections being made
- Less equipment downtime, due to better equipment prediction
- Provision of better customer service to occupants
- Extended life of facility equipment, systems, and components, resulting in dollar savings to the organization
- Improved operation and condition of the facilities

Chapter

3

Condition Assessment Customer Survey and Find/Fix Program

This chapter provides an example of a customer survey that can be used to solicit information. There are many different ways of obtaining information . . . this is just one!

Example

In an effort to evaluate and improve our service to the organization, we are in the process of evaluating our many recurring maintenance and repair programs. Our programs are numerous, broad in scope, and designed to ensure the safety and reliability of buildings, equipment, and systems. Many of the systems maintained are completely invisible until there is an emergency such as a fire or a power failure. You may have experienced the Facility Department's testing of various systems such as fire alarms or running emergency generators, but you probably have not had much interaction with the personnel doing the work, and you may not notice them at all.

The Facility Department programs are designed to deal with all aspects of our buildings and with you, the building occupants and users. The programs are operated by multidisciplinary teams of tradespersons tasked with finding and fixing deficiencies within the team's on-site capabilities. The trades teams also have the responsibility of reporting to their supervisors deficiencies that will require more than two (2) man hours to correct. The theory behind this approach is that the teams will find many minor deficiencies and fix them immediately. This will eliminate a call to the work control center and reduce the number of trips per corrected deficiency to be made to your facility. The reporting of potentially larger deficiencies will initiate inspections and analyses to identify hidden problems before they render buildings or systems inoperable or unsafe. The trade teams complete checklists showing all work performed. The lists are reviewed to determine if there are patterns that would indicate hidden problems. The analyses of checklists and referrals are geared to separate symptoms from problems so that corrective actions can be focused appropriately.

To date our evaluation of the program has been based solely on our opinions of the condition of the buildings, equipment, and systems and on how efficiently the program meets the established goals and standards. The program has been adjusted and fine-tuned as much as possible without feedback from you. We now need your views on the role the Find/Fix Program plays in the maintenance of our facilities. Please complete the attached survey and return it the Facilities Department addressed to the "F/F Survey."

1. Please identify the building in which you spend the majority of your time. Use the scale below to rate the condition of the listed areas and components of the building.

Building ____________________

5	4	3	2	1
Excellent Condition	Good Condition	Acceptable Condition	Fair Condition	Poor Condition

Main entrance: _____
Stairwells: _____

Hallways: ____
Restrooms: ____
Your office: ____
Labs: ____
Lighting: ____
Doors/locks: ____
Windows: ____
Heating/ventilation/air conditioning ____
Other area or component not on the above list: ____
(Please identify) ____________________: ____

2. Are you the person in your department who notifies the Facility Department of deficiencies?
   YES ____ NO ____
3. Are you aware of the Facilities Department's Find/Fix maintenance program?
   YES ____ NO ____
4. If you are aware of the program, are you aware of the times the Find/Fix program is scheduled for your building?
   YES ____ NO ____
5. If you are aware of when the Find/Fix teams are scheduled, do you prepare for the visit by collecting a list of deficiencies for the teams to address when they are in the building?
   YES ____ NO ____
6. If you have dealt with the Find/Fix crews, have they been courteous and helpful?
   YES ____ NO ____
7. Have you ever seen one of the Find/Fix door knob tags left behind by crews in your building?
   YES ____ NO ____
8. Have you ever returned one of the doorknob tags to the Facilities Department with comments or praise?
   YES ____ NO ____
9. Do you think that the Find/Fix program is effective in dealing with minor building deficiencies that you other-

wise would have to call in to the Facilities Department Work Control Center?
YES _____ NO _____

10. Have the number of minor maintenance items with which you must deal with increased or decreased in the past year (check the appropriate designation)?
Increased _____ Decreased _____

11. Historically the Find/Fix program schedule has included three cycles of visits to all buildings sometime during the August to April timeframe. Based on your experience with the program and the condition of your building, indicate how you would change the frequency of Find/Fix visits in your building. Circle one.

5	4	3	2	1
Four Cycles	Three Cycles	Two Cycles	One Cycles	Stop Find/Fix

12. Do you think that the Facility Department is sensitive to your needs as a building user?
YES _____ NO _____

13. If you think that the Facility Department is sensitive to your needs, do you think that the Find/Fix program represents this sensitivity?
YES _____ NO _____

14. Not only do we need your feedback using this survey concerning the Find/Fix program, we also need your suggestions concerning what you think we should do. This is an opportunity to let us know what you think about an area not covered by this survey.

If you would like to schedule a meeting to discuss the Find/Fix program or anything else related to the Facility Department or the condition of your building, please list your name, department, and phone number. After we have analyzed the results of this survey, we will call and schedule a time to discuss your concerns.

Thanks for your help!

PART 2

Assessment Inspection Checklists (Detailed)

Division

I

Building Systems Assessment

E	U	R	D

I. *OBJECTIVE:* The basic objective is to maintain buildings in a safe, economical manner that will be consistent with functional/mission requirements, sound architectural and engineering practice, and reasonable appearance.

II. *DEFINITIONS:*

A. Buildings. Buildings of all types are included, such as: administration, education, laboratories, research, operations, computer, warehouse, maintenance, radar, training, test, storage, assembly, aircraft support, antenna, sewage treatment plant, prison, model shops, utilities, photo laboratory, trailers, and incinerator.

B. Sections. For purposes of Facilities Condition Inspections, buildings are divided into the following Building System Assessment for which inspection checklists are established.

1. Chimneys and stacks

E	U	R	D

2. Buildings/structures (except roofs and trusses)
3. Roofs
4. Trusses
5. Trailers
6. Air conditioning systems
7. Cranes and hoists
8. Elevators, platform lifts, dumbwaiters
9. Food preparation and service equipment
10. Heating systems
11. Plumbing systems
12. Ventilation and exhaust air systems
13. Hot water systems
14. Electrical systems
15. Lighting
16. Switchgear

III. *MAINTENANCE STANDARDS:* Exterior and interior finishes should be maintained to correct all defects or damage in order to keep the buildings in good safe, sanitary, and operational condition. Materials or equipment worn or damaged beyond economical repair, or substandard materials, or items of equipment, may be replaced with more durable materials or equipment, provided the increased cost of such work is warranted by the proposed future use of the facility.

IV. *SCORING OF INSPECTION CHECKLISTS:* see Chapter 2 for details. Ratings are: *Emergency* (*E*), *Urgent* (*U*), *Routine* (*R*), *Deferred* (*D*).

Section 1. Chimneys and Stacks

1. *FOUNDATIONS:* Settlement; cracks due to heat, shocks, vibrations.
2. *BRICK AND CONCRETE WALLS:* Weathering, cracking; spalling; eroding or sandy

E	U	R	D

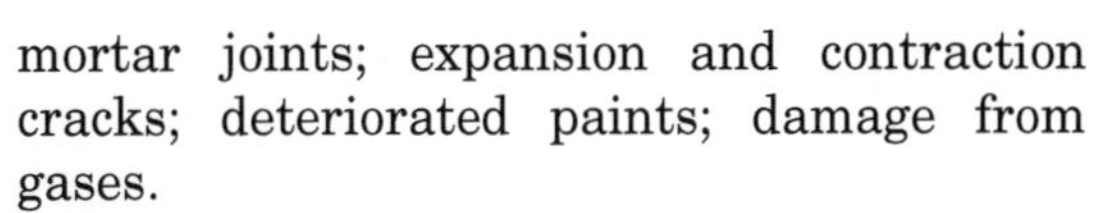
mortar joints; expansion and contraction cracks; deteriorated paints; damage from gases.

3. *CAPS:* Weathering, cracking; spalling; loose material.
4. *EXPOSED METAL SURFACES:* Rust, corrosion, and deteriorated paint, broken, loose, missing, other damage to bolts, rivets and welds.
5. *LININGS, SUPPORTING CABLES, AND BAFFLES:* Cracks, spalling; damage from gases.
6. *GUYS, ANCHORAGE, AND BANDS:* Lack of tautness; rust; corrosion: frayed, broken, loose, missing, mechanically damaged anchorage.
7. *LADDERS:* Rust; corrosion; paint sealing; poor anchorage; broken, loose, or missing ladder rungs.
8. *PAINTERS TROLLEY:* Wear, corrosion, other damage to tiller ropes, pulleys, and pulley supports; poor pulley support anchorage.
9. *OPENINGS FOR CLEANOUT DOORS, BREECHINGS, AND FLUES:* Cracking or spalling of the masonry surfaces; metal frames for distortion, rust, corrosion; broken, loose, missing, other damage to bolts, rivets, and welds.
10. *CLEANOUT DOORS AND FASTENINGS:* Distortion; rust; corrosion; wear; broken, cracked and missing parts; other damage.
11. *SPARK ARRESTER SCREENS:* Clogged with fly ash; rust, tears, and other damage; bolts and screws for rust, corrosion, loose, broken, or missing parts.
12. *LIGHTNING RODS, TERMINALS, CABLES, AND GROUND CONNECTIONS:* Corrosion; loose; burned; missing; or other damaged parts and connections. Test for elec-

E	U	R	D

trical continuity from aerial terminals through ground connections.

13. *LIGHTS, HOODS, REFLECTIONS, SHIELDS, AND RECEPTACLE FITTINGS:* Failure to operate; missing, loose or damaged parts; moisture; need for relamping.
14. *CONDUIT:* Breaks and other damage. Remove conduit inspection plates and examine internal connections for lack of tightness and inadequacy. Check relays for defective operation and for loose or weak contact springs; worn or pitted contacts and moisture.

Total Items: 14 Raw Total %

Findings (comment on each checkpoint and summarize to justify scoring):

___

___

___

___

___

___

___

___

___

___

___

E	U	R	D

Section 2. Buildings/Structures (Except Roofs and Trusses)

SCOPE: Building foundations, walls, floors and floor coverings, and roof framing and sheathing. Roofs and trusses are covered by other checkpoints.

1. *FOUNDATIONS:*
   a. Concrete: Spalling, broken areas; leaks and dampness; exposed reinforcing; out of plumb; differential settlement or frost leaks indicated by cracks, bending of doors and windows, and/or separation of wall or slab from footing.
   b. Masonry: Eroded or sandy mortar joints; mortar cracking and pulling away from brick; soft or spalling brick; leaks and dampness; out of plumb; differential settlement or frost leaks indicated by cracking, warping of doors and windows, and/or separation of wall or slab from footing.
   c. Timber: Warping; checking; splitting; bowing; sagging or broken members; deflection, rotting, termite or other insect infestation; fungus growth; evidence of dampness of long duration; damage or possible damage to wooden parts due to closeness (within 8 inches) or direct contact with soil; loose, damaged, or missing bolts, split rings, other connections.
   d. Vent screens: Binding, jamming, poor fit of frame, missing hardware or connections;

E	U	R	D

rust, corrosion of metal parts, wooden parts for rotting or other damage; blocked or covered vent openings in poorly fitted frames; loose, broken, or missing hardware or connections; metal parts for holes or rotting fabric; and other damage; clogged, blocked, or covered hole openings or inadequacy of ventilated area in floor condensation.

e. *Termites:* Termite tubes on or penetrating any masonry, concrete block, or semiporous foundation; form boards, scrapwork, or other cellulose materials under buildings and/or near foundation or structural wood.

f. *Drainage:* Failure to connect downspouts to available storm sewers, or terminate on properly installed splash blocks; improper surface grading around structure or trash debris, or other accumulations resulting in water ponding or surface runoff drainage back toward building.

2. *PAINTING: Painting:* Alligatoring; checking; blistering; crawling; cracking; sealing; peeling; wrinkling; flaking; fading; loss of gloss; excessive chalking; mildew; bleeding; staining caused by screens or splashing; discoloration; complete absence of paint. Repaint when paint film is less than 3 mils in thickness on metal surfaces. Painting of concrete is not recommended when required for damp-proofing.
3. *WALLS—EXTERIOR:*
   a. *Concrete:* Cracks; spalling; broken areas.
   b. *Masonry:* Eroded or sandy mortar joints; mortar cracking and pulling away from brick; soft or spalling brick; leaks and dampness; out of plumb; differential settlement or frost leaks indicated by cracking, bending of doors and windows, and/or separation of wall or slab from footing.

E	U	R	D

c. *Parapet walls:* Loose coping stones; eroded or sandy mortar joints; expansion cracks.

d. *Stucco:* Corrosion, disintegration; alligator cracks; water stains; broken areas.

e. *Cement-Asbestos:* Stains; loose fastenings; broken sheets or shingles.

f. *Aluminum:* Corrosion from corrosive atmosphere; loose fastenings; other damage.

g. *Ferrous and other metal:* Scars; scuffs; rust; corrosion; breaks in protective coatings; puncture of metal; loose, missing, or broken fastenings.

h. *Wood:* Looseness; warping; surface cracking; knots; rotting; fungus or termite infestation; stains; evidence of continual and other wooden parts for ventilation and condensation infestation; damage or rot from ground level.

i. *Weathertightness:* Lack of weathertightness around framed openings, where butting dissimilar surfaces between joints.

4. *WALLS—INTERIOR AND CEILINGS:*

a. *Wood:* Checking, cracking, splintered, broken; warping; sagging; support failure; rot; termite or other insect or fungus infestation; abrasion; scuff marks; mechanical damage; personal abuse.

b. *Soft fiberboard (acoustical and insulation):* Open joints; buckling; sagging; support failure; loose; missing; failure of fastenings or adhesive; abrasions; breaks; holes; stains from weather or utility leaks.

c. *Wallboard, plasterboard, hard-pressed fiberboard and cement-asbestos board:* Open joints; cracking; buckling; sagging; support failure; loose: failure of nailing or adhesive; abrasion; breaks; holes; discoloration from weather or utility leaks.

d. *Plaster:* Cracking, buckling, sagging, support failure; spalling, crumbling, or falling

E	U	R	D

from moisture absorption; efflorescence, peeling, or flaking from moisture or sealer failure; discoloration from weather or utility leaks.

e. Ceramic tile: Chipped, cracked, loose, missing, holes; defective mortar joints; etched, pitted, or dull surfaces caused by use of acidic or abrasive cleaners.

f. Metal: Corrosion, rust, abrasions, indentations, punctures, deterioration of protective coating.

5. *WALL COVERINGS:*

a. Resilient coverings (linoleum, vinyl, plastic): Curling, loose, adhesive failure, abrasions, indentations, punctures, tears; etched, pitted, or dull surfaces caused by the use of acidic or abrasive cleaners.

b. Canvas and paper: Breaks, wrinkling, fading, adhesive failure.

c. Weather or utilities leaks: Damage to wall coverings, evidence of mechanical damage or personal abuse.

6. *DOORS AND WINDOWS:*

a. Wood sash, doors and trim: Splitting, rotting, cracking, loose, poor fit, binding, missing, loose or missing caulking, lack of weathertightness.

b. Metal sash and doors: Rust, corrosion, warping, binding, poor fit, lack of weathertightness.

c. Storm sash: Metal parts rotting and binding, jamming, poor fit of frames; rust and corrosion; wood parts damaged.

d. Screens: Loose, broken, or binding, jamming, poor fit, rust, corrosion, holes in screen from rotting, stain, other damage; poor fit of frames; wood parts rotting, stained, other damage.

e. Shutters: Splitting, rotting, cracking, loose, missing, misalignment, and freedom of

E	U	R	D

spring, or little or no free motion as required.

f. Hardware: Loose, missing, broken parts; binding, misaligned, improper installation or adjustment, lack of lubrication; corrosion, abrasion, loss of finish coating.

g. Glass: Missing or broken panes; disintegration of putty.

h. Trim: Looseness, scratches, indentations, mechanical damage, personal abuse.

i. Venetian blinds, window shades: Insecure or broken fasteners; poor operation; frayed or broken cords or tapes; broken or damaged slats; worn or torn material.

7. *FLOORS AND STAIRS:*

a. General: Inadequacy of floor load postings or lack of conformance by occupant to posted loadings; lack of rigidity of supporting beams or other structural supports and need for immediate correction; loose, splintered, missing, or other damage to handrails, lattices, and supports.

b. Wood: Sagging, wear, splintered, loose, warped, scratched, shrinkage, cracks, rotted, stained, discolorations, indentations, moisture penetration, broken, absence of protective coatings, evidence of insect infestation.

c. Concrete: Wear, pitting, roughness, discolorations, stains, settlement, shrinkage cracks, particularly where placed over wood framing; absence of treatment or waxing that would prevent surface dusting; treads and risers for wear, cracked, chipped, or damage caused by settlement.

d. Mastic: Wear, depressions, indentations, other damage resulting in uneven surfaces; absence of wax protective coatings.

e. Terrazzo: Cleanliness, wear, pitting, roughness, discolorations, settlement, cracks.

E	U	R	D

f. *Magnesite* (*magnesium oxychloride*): Wear, chipping, roughening; settlement cracks; failure due to inadequate minimum thickness (5/8 inches); and damage from exposure to moisture.

g. *Clay and masonry tile:* Sandy and eroded mortar joints; individually stained, or broken, chipped, or loose, resulting in uneven surfaces.

h. *Steel:* Wear, rusted, loose, bent, or other damage to surfaces, risers, treads, and structural supports; broken welds and loose, missing or damage bolts, nuts, rivets, and screws.

8. *FLOOR COVERINGS:*

a. *Carpets and rugs:* Wear, tear, cuts, raveling, discolorations, fading; binding or anchoring strips for wear, damage, poor anchorage; worn or missing tractive substance on backing of small rugs or carpets where placed over polished floors; beetle or moth damage.

b. *Resilient floor coverings* (*linoleum, vinyl, plastic, vinyl asbestos, cork, rubber, and asphalt tiles*): Wear, cracking, chipping, breaking, scratches, tears, indentations, lack of bonding and unevenness of underlayment; evidence of dramage resulting from use of solvents or excessive use of water for cleaning; absence of protective wax coatings.

c. *General:* Loose, damaged, or missing bases, binding strips, or thresholds; projecting nails, bolts or screws; slippery surfaces from oil or other spillage; inadequate exposure of abrasive necessary for nonslip finish surfaces; loose, missing, broken, or other damage to abrasive stair risers or treads; lack of nonslip finish in other types of stair coverings.

E	U	R	D

9. *ROOF FRAMING:*
   a. *Roof rafters, joists, and sheathing:* Checks, splits, broken members, open joints, loose boards, sag of members, displacement of joints, insect damage.
   b. *Gutters and downspouts:* Misalignment, rust, corrosion, material accumulations and clogging, breaks, leaks, missing wire guards, loose or missing hangers or other fastenings.

Total Items: 9 Raw Total %

Findings (comment on each checkpoint and summarize to justify scoring):

____

____

____

____

____

____

____

____

____

____

____

____

E	U	R	D

____

____

Section 3. Roofs

SCOPE. Built-up, asphalt shingle, roll-roofing, cement-asbestos, metal, slate, tile, concrete slab, and wood-shingle roofs.

1. *WOOD-SHINGLES:* Wear from weathering; warped, broken, split; excessive curling; missing; flashing failures.
2. *TILE:* Wear from weathering, cracked, loose, missing, flashing failures, deterioration of expansion-joint material or tile arising from improper placing or inadequate expansion joints.
3. *SLATE:* Wear from weathering; broken, cracked, loose, missing; flashing failures.
4. *METAL:* Holes, looseness, punctures, broken seams, inadequate side and end legs, inadequate expansion joints, rust or corrosion, damage resulting from contact of dissimilar metals.
5. *CEMENT-ASBESTOS:* Wear from weathering, broken, cracked, loose, missing, insufficient side or end lap.
6. *ASPHALT ROLLS:* Weathering, cracking, alligatoring, buckling, blistering, insufficient or uncemented laps, tearing from nails too close to edge, other damage to coatings.
7. *ASPHALT SHINGLE:* Lifting, weathering, cracking, curling, buckling, blistering, loss of granules, excessive exposure.
8. *BUILT-UP:* Cracking, alligatoring, low spots, and water ponding; failure or lack of gravel stops; cracks in membrane; exposed bitumi-

E	U	R	D

nous coatings; exposed, disintegrated, blistered, curled, or buckled felts.

9. *FASTENINGS:* Improper materials, loose, missing, broken, defective, exposure.
10. *METAL BASE FLASHINGS:* Rust, open vertical joint; loose flanges; inadequate or exposed nailing; improper fastening; improper sealing with felt strips; deteriorated or missing cant strip.
11. *OTHER BASE FLASHINGS:* Sagging, separation; inadequate coverage or embedment; open vertical joints; improper fastening; buckling; cracking; surface coat disintegration; deteriorated or missing cant strip.
12. *METAL CAP FLASHINGS:* Rust, corrosion, open joints, loose, improper fastenings.
13. *OTHER CAP FLASHINGS:* Open joints, buckling; cracking; surface coat disintegration; improper fastenings.
14. *CHIMNEY, WALL, RIDGE, VENT, VALLEY, AND EDGE FLASHINGS:* Open joints, loose, improper fastenings, other damage.
15. *CONCRETE SLABS:* Cracks, spalling, expansion joint deterioration, low spots, improper drainage.
16. *PARAPET WALLS AND COPINGS:* Cracks, spalling, defective joints, other damage.
17. *GENERAL:* Need for rust removal and painting or protective coating of exposed metal surfaces.

Total Items: 17 Raw Total %

Findings (comment on each checkpoint and summarize to justify):

____

____

E	U	R	D

Section 4. Trusses

SCOPE. All types of building trusses that are normally designed for the express purpose of supporting roof loads as well as ceilings where applicable. All lateral and vertical bracing and ties between trusses are included.

1. *TIMBER:*
   a. From ground: Twisted and bowed members; excessive number and size of knots; slope of grain over 1 inch in 10; checks and splits in ends of web member; separation or slippage at joints; sag; overloading.

E	U	R	D

b. From truss: Loose bolts, split rings, shear plates, and fastening devices; checks and splits in bracing, chord member, splice plates (scabs), web member and filter blocks; missing filler blocks; improper end and edge distances; looseness of the rods (bolts may be considered loose if after head of bolt is struck a sharp blow with a hammer, the nuts can be taken up two full turns or more.

c. *Steel splice plates:* Rupture, shearing, crushing, rust.

2. *STEEL:*

a. *From ground:* Twisted, bowed, deformed, broken members.

b. *From truss:* Loose bolts, nuts, defective welds, rupture, shearing, or crushing of steel plates, members, bolts, and rivets.

3. *TRUSS SAG:*

a. *General:* Examine trusses for sagging by stretching piano wire between supports and measuring vertical deflections at panel points, or use a surveying instrument.

b. *Truss member identification:* Use a sketch of the truss and label the panel points either numerical or alphabetically. The deficient member may then be identified by indicating the panel points between which it lies.

4. *DRY ROT, TERMITES, AND OTHER INFESTATION:*

a. *Wooden parts:* Dampness and surface moisture of long duration; termite; other insect, and fungus infestation. Termite and fungus infestations are often detected prior to actual visual damage by probing with a sharp pointed instrument in those areas where prolonged dampness is not directly associated with rainfall or damp climate.

E	U	R	D

b. Wooden supports near or at ground level: Termite tubes or tunnels; dirt piled up to wood level; need of protective treatment. If removal of dirt pile is impractical, wood in direct contact with, or less than 8 inches from ground level should be given protective treatment.

5. *PAINT: Painted surfaces:* Blistering, checking, cracking, sealing, wrinkling, flaking, mildew, bleeding rust, corrosion; complete absence of paint, particularly at ends of members; record film thickness and condition on metal surfaces.

Total Items: 5 Raw Total %

Findings (comment on each checkpoint and summarize to justify scoring):

____

____

____

____

____

____

____

____

____

____

E	U	R	D

Section 5. Trailers

SCOPE: Trailer roofs, walls, floors and floor coverings and stairs.

1. *ROOFS:* see SECTION 3: *ROOFS,* checkpoints numbered 4, 6, 8, 9, and 14.
2. *BUILDINGS:* see *BUILDINGS/STRUCTURES CHECKPOINTS (EXCEPT ROOFS AND TRUSSES)*, checkpoint numbers:
   Painting—2.
   Walls Exterior—3. f, h, and i
   Walls—Interior and Ceilings—4. a, b, c, and f
   Wall Coverings—5. a and c
   Doors and Windows—6. b, d, f, g, and h
   Floors and Stairs—7. a and b
   Floor Coverings—8. a, b and c

Total Items: 2 Raw Total %

Findings (comment on each checkpoint and summarize to justify scoring):

____

____

____

____

Section 6. Air-Conditioning Systems

SCOPE: Air-conditioning, ventilating, and related equipment, including refrigeration units, evaporative cooler, fans, or other equipment, shall be maintained at a level consistent with the service required.

1. *AIR-CONDITIONING, WINDOW UNIT:*

E	U	R	D

a. Check condenser, cooling coil fins, and fans for cleanliness.
b. Check all interior parts for dirt or dust.
c. Check filter for cleanliness.
d. Inspect damper for correct setting.
e. Check motor and fan bearings for lubrication.
f. Inspect gaskets for leaks.
g. Check for refrigerant leaks.
h. Check frame of unit with ohmmeter for proper electric ground.
i. *Start unit controls.* Check temperature differential and controls.

2. *AIR-CONDITIONING MACHINE PACKAGE UNIT (COMFORT COOLING):*
   a. Inspect interior and exterior of machine for cleanliness.
   b. Check drain pan for cleanliness and corrosion.
   c. Check for refrigerant leaks.
   d. Check refrigerant levels.
   e. Check condition of cooling and reheat coils.
   f. Check belts for wear, tension, and alignment.
   g. Inspect humidifier drip pan for cleanliness and corrosion.
   h. Check alignment of motor and fan.
   i. Check prefilters and final filters for condition.
   j. Check compressor oil level.
   k. Check and adjust vibration eliminators.
   l. Run machine, check action of controls, relays, switches, etc.
3. *AIR-CONDITIONING, MACHINE PACKAGE UNIT (COMPUTER ROOM):*
   a. Inspect interior and exterior of machine for cleanliness.
   b. Inspect drain pan for cleanliness and corrosion.

	E	U	R	D
c. Check for refrigerant leaks.				
d. Check refrigerant levels.				
e. Check condition of cooling and reheat coils.				
f. Inspect humidifier drip pan for cleanliness and corrosion.				
g. Check motor and fan bearings for lubrication and alignment of motor and fan.				
h. Check belt tension and condition.				
i. Check prefilters and final filters for condition.				
j. Check compressor oil level.				
k. Check and adjust vibration eliminators.				
l. Run machine, check action of controls, relays, switches, etc.				
4. *COOLING TOWER:*				
a. Inspect wet deck for cleanliness.				
b. Inspect protective finish looking for spot corrosion.				
c. Check gear box oil.				
d. Check structure for deterioration.				
e. Inspect motor, belts, etc. for proper operation.				
f. Check alignment of gear, motor, and fan.				
g. Check eliminator for buildup.				
h. Check water distribution.				
i. Check fans and air inlet screens.				
j. Check nozzles for clogging and proper distribution.				
k. Inspect keys and keyways in motor and drive shaft				
5. *EVAPORATIVE CONDENSER—UP TO 50 TONS / OVER 50 TONS:*				
a. Inspect water pans for dirt, trash, and algae.				
b. Check water outlets and coil connections.				
c. Check oil in gear reducer, fan and pump.				
d. Check gear box, bearings, alignment, etc.				
e. Check drive shafts.				
f. Check control and float valves.				

E	U	R	D

g. Inspect eliminators.
h. Inspect condenser coil, fins, sprays, connections, etc.
i. Check screen.
j. Check water treatment equipment.
k. Check motors and starters.
l. Check structural fittings.
m. Determine if continuous bleed line is open.
n. Determine if unit is cleaned from a recent drain and flush out.

6. *REFRIGERATION MACHINE (ABSORPTION)*:

a. Check unit strainer and unit pump motor cooling circuit for cleanliness.
b. Check for cleanliness all strainers and traps in steam or hot water supply, condensate return, and condensing water circuit.
c. For unit with extended purge pump system, check pulley alignment and U-belt extension.
d. Check machine oil level in purge motor pump.
e. Check service control systems.
f. Inspect cooling water circuit.
g. Inspect system water circuit. Check log sheet for indications of increased temperature trends.
h. Check pumps, motors, and controls (evaporator pumps solution pump).

7. *REFRIGERATION MACHINE (CENTRIFUGAL AND RECIPROCATING)*:

a. Check lubrication for drive couplings, motor bearings (nonhermetic), vane control linkage bearings, ball joints, and pivot points, compressor oil reservoir, purge compressor, and gearbox.
b. Check alignment of line couplings.
c. Inspect cooler and condenser tubes for scale.

E	U	R	D

d. Inspect purge unit.
e. Check water strainers for cleanliness.
f. Check refrigerant level.

8. *AIR-CONDITIONER, UNITARY ROOF-TOP (HEATING AND COOLING)*:
   a. Check for debris on air server and underneath unit.
   b. Inspect gasketing for leaks.
   c. Check condenser, cooling coil fins, and fans for cleanliness
   d. Check filter for cleanliness.
   e. Inspect and adjust damper.
   f. Check motor and fan bearings for lubrication.
   g. Check bearing caller set screws on fan shaft to make they are tight.
   h. Check dampers for dirt accumulation.
   i. Check damper motors and linkage for proper operation.
   j. Check coils for cleanliness and leakage.
   k. Check belts for wear, adjust tension or alignment.
   l. Check rigid couplings for alignment on direct drives for tightness of assembly. Check flexible couplings alignment and wear.
   m. Check electrical connections and mounting for tightness.
   n. Check for corrosion.
   o. Check mounting bolts for tightness.
   p. Check vibration eliminators for proper adjustment.
9. *COMPRESSOR:*
   a. Check compressor oil level.
   b. Run machine, check action of controls, relays, switches, etc. to see that:
      1. compressor(s) run at proper settings
      2. reheat coils activate properly
      3. crankcase heater runs properly
      4. section and discharge pressures are proper

	E	U	R	D
5. discharge air temperature is set properly				
10. *HEATING UNIT (GAS AND OIL FUELED)* (if equipped):				
a. Check burner for flashback and tight shut off of fuel				
b. Check operation of controls.				
c. Check burner, chamber, thermocouple, and control for cleanliness.				
d. Check pilot or ignition device for proper setting.				
e. Inspect vent and damper operation.				
f. Check operation of safety pilot, gas shut-off valve, and other burner safety devices.				
g. Check burner adjustment while unit is operational.				
h. Check temperature differential controls.				
11. *HEATING UNIT (ELECTRICAL)* (if equipped):				
a. Inspect for broken parts, contact arcing, or any evidence of overheating. Inspect all wiring for deterioration.				
b. Check name plate for current rating and controller manufacturer's recommended heater size.				
c. Check line and load connections and heater mounting screws for tightness.				
Total Items: 11 Raw Total %				
Findings (comment on each checkpoint and summarize to justify scoring):				
____				
____				
____				

E	U	R	D

___

___

___

Section 7. Cranes and Hoists

SCOPE: Cranes and hoists, including overhead traveling, wall gantry, jib, pillar and pillar jib cranes, monorail hoists, and simple electric hoists.

1. *HOUSEKEEPING:* Oil and grease spillage, accumulated drippage, litter, trash, loose boxes, pails, rags; recesses or pocket trapping or accumulating, water, dirty, illegible, or missing signs, labels, instructions, or nameplates.
2. *LUBRICATION:* Overflow of oil or grease from antifriction bearings of over 800 rpm; plugged or frozen grease fittings, ineffective pressure relief fittings; oil leaks, foaming, improper oil level in bearing boxes or gear casings; worn or damaged gaskets or seals; other than thin film on grease-lubricated open gearing. Check residue in pan that contacts gears.
3. *STRUCTURAL FRAMING:* Bends, dents, cuts, rope-rubbing points, corrosion, loose rivets and bolts evidenced from observation or by hitting lightly with hammer; cracks, particularly around rivet and bolt holes, at sharp re-entrant angles, through throat of molds and along edge, particularly at ends of girders, trucks, and near broken sections; overstrain detected by flaking of paint along lines at 45° angle to axis of member or in broad bands across one side of member

E	U	R	D

(likely to occur at re-entrant angel cutouts and loaded ends of stiffness).

4. *BRIDGE END AND GANTRY TRUCKS:* Nonintegrity of diaphragms and diaphragm connections, wear of rocker pinholes, insecure rocker pins, loose wheel-bearing boxes, loose bolts in aligning gussets, corrosion.
5. *CABLES:* Warped or misfitting broken sections; overstrain detected by flaking of paint along lines at 45° angle to axis of member or in broad bands across one side of member (likely to occur at re-entrant angle cutouts and loaded ends of stiffness).
6. *PLATFORMS, LADDERS, AND STAIRS:* Loose and rotted planking and platform edge openings, open and inoperative trap doors, bent handrails, loose or bent ladder rings and rails, smooth worn spots on platforms and stairs, defects and improper operation of gate hardware and springs, corrosion.
7. *TROLLEY RAILS:* Roughness, peeling, rolling over of the heads, cracks in heads, gaps between joints, loose bolts or clips, longitudinal slippage of wearing and bearing surfaces, corrosion.
8. *COMBINATION BRIDGE AND TRACK MEMBER:* Rough, peeled, rolled, and cracked trash flanges; wavy and uneven wheelbearing surfaces; worn wheel-guide surfaces; damage to joints, corrosion.
9. *CONDUCTOR SUPPORTS:* Loose or improperly adjusted, resulting in possible improper tracking of collectors.
10. MONORAIL SWITCHES: Excessive width of gaps, out of level, misalignment of joints; frayed or worn pull ropes or chains; damaged or rough operating handles; worn pivots and sliding surfaces; evidence of pounding; ineffectiveness and deterioration of baffles, locks,

E	U	R	D

guides, and interlocks; wear of springs and attachments, corrosion.

11. *BUMPERS AND ENDS STOPS:* Loose bolts or rivets; cracks, cracked welds; out-of-line contact, evidence of longitudinal slipping; split wooden contact blocks; general deterioration from excessive pounding.
12. RAIL WIPER AND BRUSHES: Wear on sides of wiper extending over sides of railhead indicates misalignment of crane, also evidenced by wear of wheel flanges; wiper out of place and needing realigning to clear rail; wear or severe tangling of bristles of brushes.
13. *GUARDS:* Looseness, misplacement, ineffectiveness; damaged or worn hardware, defective gaskets. Close and secure gear guard openings; replace defective gaskets.
14. *CABINETS:* Corrosion, particularly in bottoms and cones of interiors; clogged vents; poor fit of doors; damage to hardware, clear vents.
15. *GEARS:* Looseness of shaft; teeth are dull, rounded, chipped, broken, cracked.
16. *SHAFTS:* Bends, flutter in action, cuts and nicks, loose or worn keyways, misalignment, corrosion.
17. *BEARINGS:* Looseness, chatter, excessive heat of sleeve bearings; worn shields, noise, beating of antifriction bearings, visible discolorations indicate overheating from overloading, misalignment, or improper lubrication practices; uneven loading on bearing, wear of pins, pinholes, untrue vertical placement that may cause jib to drift laterally for pintle bearings of jib, pillar, and pillar jib wear.
18. *BRAKES (GENERAL):* Excessive wear in linkages, pins, cams; weakness of springs; wear and general deterioration of linings;

E	U	R	D

roughness of drum; clearance between drum or dirks not in accordance with manufacturer's instructions; toggle points approaching dead center too closely and no allowance for lining wear when brake is applied. Adjust brake if pedal travel is approaching its ultimate under full pedal pressure.

19. *HYDRAULIC BRAKES AND HYDRAULIC ELECTRIC BRAKES WITH HYDRAULIC-LOCKING MECHANISM OR SOLENOIDS:* Leakage indicated in lines to wheel or to master cylinder if pedal gradually moves with contact foot pressure; overtravel of wheel cylinder indicates improper lining and drum clearances; softness felt in pedal action indicates air in system; malfunctioning of control unit, push buttons, and signal lights of locking mechanism; improper solenoid air gap; evidence of overheating; damaged brass air gap material and loose core laminations, particularly on AC brakes; delay or restriction in opening of brakes; adjusted length of holding spring not in accordance with manufacturer's set length. Adjust to proper lining and drum clearances; fill reservoir when necessary; bleed air from system where soft pedal action and refill reservoir; correct maladjustment of holding spring.
20. *ELECTRIC SOLENOID BRAKES:* Improper solenoid air gap; evidence of overheating; damaged brass air gap material and loose core laminations, particularly on AC brakes; delay or restriction in opening of brake; adjusted length of holding spring not in accordance with manufacturer's set length. Correct maladjustment of holding spring.
21. *THRUSTER BRAKES:* Low level and improper consistency of hydraulic fluid, dirt, looseness of orifice settings, looseness of hold-

E	U	R	D

ing spring setting; deterioration of motor; worn brakes, burned or high commutators of DC motors; loose bearings, misfitting brakes; rough or dull slip rings on AC motors.

22. *EDDY CEMENT BRAKES:* Loose bearings, signs of dragging armature, looseness of anchor bolts.
23. *ROPE DRUMS AND SHEAVES:* Cracks, particularly around hubs; chips, wire corrugations, other forms of roughness on rope grooves and flanges.
24. *AUTOMATIC MECHANICAL LOAD-LOWERING BRAKES:* Rough, unpolished screw threads, cams, and friction plates; worn or damaged friction linings; obstructed oil passages in linings; worn, peeled, or chipped ratchets and pawls; excessive clearance between friction plates, between ratchets and pawls, and between restraining band and band drums; too-heavy viscosity and loss of lubricating properties. Defective operation noted when lowering loads, such as irregular speeds, jerky lowering, and jerky starting from rest; indicate defects resulting from one or more of the above conditions.
25. *HOOKS:* Rough or sharp surfaces for rope slings, cracks, or deformations from true shape, interference to smooth and easy swiveling, slings.
26. *WIRE ROPES:* Slippage, loose nuts, bolts, wedges, improperly placed clips for all connections; broken wires, internal corrosion, kinks, high strands, excessive wear (when one thread of cross-sectional area of outer wires is worn away), note causes of crushed wires, popped cores, and birdcages; internal corrosion within and between strands of standing parts and portions adjacent to sockets for broken wires; lack of thick, clinging

E	U	R	D

preservative material on internal and external strands (use extreme caution in separating wires, to prevent damage).

27. *WHEELS:* Uneven wear, flat spots, sprawls, chips, (particularly on wheels with tapered tracks); pattern of wear on flanges for possible misalignment of wear and loss of reversal on tracks; cracks, especially in flanges of cast iron wheels; looseness of axle pins or securing devices, misplacement of pins.
28. *COUPLINGS:* Looseness of compression type, wear of grids, teeth, and lugs of grid or internal types; lack of lubricant, worn or disintegrated seals; rotational looseness and decomposition of materials of couplings with rubber, plastic, or composition parts.
29. *MECHANICAL LOCKS AND INTERLOCKS:* Irregularities of action, worn or out-of-shape latch sockets, bent shear brass, tendencies to release under load, insecure setting, and misalignment.
30. *ELECTRICAL CIRCUITS:* Corroded, loose, non-watertight conduit and fittings; breaks or openings in armored cable; broken cracked, or otherwise damaged insulation; loose or insecure wiring connections; signs of burns or flashburns; incomplete or inadequate grounds; accidental or inadvertent grounds. Tighten loose wiring connections; remove accidental or inadvertent grounds.
31. *BRIDGE AND RUNNING CONDUCTORS:* Insulators; conductive dust, arc tracks, other evidence of arcing or leaking, breaks, chips, loss of glaze, loose or bent supports. Rigid and enclosed type: oxidation sufficient to cause high resistance or sparking; burns, pits, bends, twists, loose or rough joints, insecure joint bonds, evidence of nonuniform bearing of collectors or undesirable rubbing

E	U	R	D

or scraping of collectors on enclosures; disintegrated or misaligned enclosures.

Tension wire type: Excessive sag or supports, low tension between supports, loose locking nuts or tension devices; burns, pits, excessive wear.

32. *CURRENT COLLECTORS, INSULATORS:* Conductor dust, on tracks, other evidence of arcing or leaking, breaks, chips, loss of glaze, loose or bent supports, broken wires or loose connections in shunts or pigtails. Collectors; irregular and nonuniform bearing and wear, improper riding on conductors, loss of graphite or other lubricant; unit bearing pressure too light or too heavy; loose support connections, loose or worn pins, worn pinholes, tight or obstructed movement of support arms; improper location of pickup-type shoes to deposit wires near center of spool indicators after passage of shoes; burned or pitted wheels; worn tread, flanges, wheel bores, or pins; rough guide track for wheel type.
33. *CABLE REELS:* Bent or dented drums, tight or loose bearings, inoperable over running clutches, insecure cable connections or supports, caked dirt or foreign matter in mechanism housing; loss of seal in watertight or dust-tight enclosures; loose cover and loss of grounding in explosion-proof reels; inadequate spring tension or improper motor conditions, excessive wear on rings and brushes, signs of sparking, grooving, or scoring, weak spring pressure.
34. *LIGHTING SYSTEM:* Ineffective circuit breakers, loose or burnt-out fuses, sluggish switches, burnt-out receptacles, broken, dirty, fogged reflectors, burnt-out bulbs. Replace burnt-out fuses and bulbs.
35. *MANUAL SWITCHES:* Loose hinges, low tension in spring contacts broken, wrenched,

E	U	R	D

or loose handles of knife switches; low and rough contact segment divisions, inadequate contact shoe pressure of faceplate switches; slowness and uncertainty of snap, excessive arcing, improper action of arcing shields or suppressor of snap-action switches; wear of cams; looseness of cams or retaining means, wear of cam rollers and pins, uneven or insecure feel of star wheels, inadequate contact spring tension, evidence of moisture within enclosing boxes of cam-type water and drum-type switches; worn or frayed rope, insecure connections of rope and handles, broken handles, weak return, spring of pendant-rope-operated switches; burns, welds, pits, corrosion, misfit, uniform contact surfaces. Refinish contact points and surfaces when damage is not severe; replace contact points where damage is severe.

36. *PENDANT PUSHBUTTON STATIONS:* Defective insulation on electrical cable, particularly at junctions to pendant box and control enclosure; looseness of weight-carrying and grounding cable, failure to support weight of station; insecure grounding connections top or bottom; material not flexible, stranded, continuous wire rope; breaks or bends in suspended case; insecure gasket and slack where watertight or explosion-proof pendant boxes are used; variable and light feel of speed notches of pushbuttons, burned or worn contacts, damaged or faded colors, especially Stop button, illegible markings; dead pilot lights.
37. *CONTACTORS, RELAYS, AND ELECTRICAL PROTECTIVE DEVICES:* Striking solenoid armatures, broken or insecure arcing shields, weak pressure-holding or closing springs of contactors, relays, and electrical

E	U	R	D

protective devices, excessive hum and chattering of AC contactors; nonuniformity of timing, insecurity of air-gap shims or washers, poor consistency of oil in oil-type dashpots, dirt in air-type dashpots, obstructions in orifices, looseness of contact in adjustable-type resistors and capacitors for relays and electrical protective devices, mechanically delayed tripping, nonsetting and weak springs of electrical protective devices.

38. *TRANSFORMERS, REACTORS, AND MAGNETIC AMPLIFIER:* Excessive hum, indicating loose laminations.
39. *RECTIFIERS:* Loss of full voltage when malfunctioning of rectifiers operated or rectifier-controlled apparatus occurs.
40. *MOTOR CONTROL PANELS PANELBOARDS:* Cracks on tracks, missing or loose bolts, insecure mounting. Power Off: improper sequence and timing of all accessible components for hoisting, lowering, forward, and reverse. (proper sequence and timing obtained from wiring diagrams posted on crane). Power On: improper sequence, lack of smoothness (improper timing), improper adjustment of power resistors, inherent characteristics of controls causing rough operation. Adjustment: note correctable deficiencies of speed, torque, regulation, and acceleration peaks from low-speed covers of hoists, speed-time, curves of trolley, and rotating and traveling motions for each point of control.
41. *POWER RESISTORS:* Loss of section from pit corrosion, burns; loose clamping rods and brass; broken or chipped insulators; loose taps and connections; warped and short-circuited elements; charred or broken insulation; hot spots or visible red heat indicating too small current capacity.

E	U	R	D

42. *LIMIT SWITCHES* (must be maintained in *Excellent* condition: Unsatisfactory operation without load, excessive draft; non-dust-tight housings for interior use; non-watertight housing for exterior use; worn, cracked or rough cams, dogs, rollers, roller pins; loose cams or other setting or operating devices; signs of dragging of any part, tight bearings, sluggish snap action, insecure fastenings; broken wires in shunts or pigtails; burned or pitted contacts; improper fit of welded contacts; poor contact spring tension; too-short, bent, battered, misaligned, or displaced paddles or arms; wear, insecure guides and rope connections, inoperative rope shears, loose counterweights on rope-guided tripping devices; block does not stop drifting before plugging device operators, and mal-operation of plugging device when limit switch fails on semimanual and manual produced packaged hoists with air control circuit limit switches.
43. *SIGNAL EQUIPMENT:* Improper functioning of bells, horns, signal lights.

Total Items: 43 Raw Total %

Findings (comment on each checkpoint and summarize to justify scoring):

____

____

____

____

____

E	U	R	D

___

___

___

Section 8. Elevators, Platform Lifts, and Dumbwaiter

SCOPE: Elevators, platform lifts, dumbwaiters, load tests not covered in checkpoints. Requires technical specialties for performance.

Precautions: When starting inspection, make certain car operating device, emergency stop switch, and inspection are in proper working order and position for inspection. Try telephone and/or emergency signal bell to ensure it is properly connected and operative. Assure that mechanic's lights on top of car are operating. See that elevator is secured from service by posting signs, guards, or barricades until inspections and tests are completed.

1. *INSPECTIONS FROM INSIDE CAR :*
   - *a. Hoistway doors and gates:* Broken glass; debris in tracks; bent, sprung, or otherwise defective frames, panels, or guides; jams in travel, or operation is restricted; worn hangers, tracks, or latches. (Check at each landing.)
   - *b. Hoistway door hydraulic door closer:* Worn, broken, loose, or missing parts; leaking valve cylinder does not contain adequate liquid. (Check at each landing.)
   - *c. Hoistway door interlock:* Worn, broken, loose, or missing part; improper operation (does not prevent operation of driving machine by normal operating device unless hoistway door is locked in "Closed" posi-

E	U	R	D

tion; does not prevent opening door from landing side unless car is within landing zone and is either stopped or being stopped). (Check at each landing.)

d. Hoistway door interlock: Door locking function improper operation. (With car far enough above or below landing so cam will not release interlock, try to open door by pulling and lifting at each landing. Door shall remain closed and locked.)

e. Hoistway door interlock: Interlocking functions improper operation. (With car at each landing, open door in normal manner. Close car door or gate. Move car operating device to "On". Car shall remain stationary.)

f. Hoistway door interlock: secondary lock: Worn, broken, loose, or missing parts; improper operation. (Car will start by normal operating device when hoistway door is not in "Closed" position; interlock compact is made before pawl is engaged in rack or door is held by secondary lock; drag link or secondary lock does not prevent door being opened more than 4 inches after door is moved to within 2 inches of full closure.) (Check at each landing.)

g. Hoistway door separate mechanical lock and electrical contact: Worn, broken, loose, or missing part; improper operation (does not prevent operation of driving machine by normal operating device unless door is in "Closed" position; does not lock door in "Closed" position; does not prevent it being opened from landing side unless car is within landing zone). (Check at each landing.)

h. Hoistway door: electric contacts: Worn, broken, loose, or missing parts; improper op-

E	U	R	D

eration (with car gate closed, car shall not start by normal operating device when hoistway door is not in "Closed" position). (Check at each landing.)

i. Hoistway door: mechanical locks: Worn, broken, loose, or missing part; improper operation (door in closed and locked position can not be opened by pulling or lifting the door). (Check at each landing.)

j. Emergency doors: Access blocked; improper operation (cannot be opened manually from hoistway side without a key, except when locked "Out of Service").

k. Clearance between car and landing thresholds: power elevators: Improper clearance (clearance between car platform sill and hoistway edge of any landing sill, or hoistway side of any vertically sliding counterweighted or counterbalanced hoistway door is less than 1/2 inch where side guides are used, or less than 3/4 inch where corner guides are used; maximum clearance is greater than 1 1/2 inches). (Check at each landing.)

l. Car floor and landing thresholds: Deterioration or damage of floor surfacing; equipment or material stored in car; landing edges of any threshold and cam platform (power elevator) not sufficiently illuminated; other unsafe condition. Floating platform out of adjustment or otherwise defective; car can be called by means of landing button at another floor with 30-pound load on platform and car gate (door) open. (Car shall not move with this weight at any position on platform.)

m. Hoistway protection at shear points: Bevel plate under landing threshold or other projection in hoistway is rough or loose; recess

	E	U	R	D
opposite car openings other than windows and doors in side of the hoistway enclosure is not filled in flush with hoistway surface; filling material in recess is not securely fastened.				
n. *Car door or gate:* Broken glass; debris in tracks; excessively worn tracks or hangers; bent or sprung frame1 panel, lattice member, or guide; door or gate jams in travel, or operation restricted; car gate electric contact inoperative. (Electric contact is inoperative when, with car door or gate open and car operating device moved to "On" position and car door or gate moved slowly towards its "Closed" position, car will start before "Closed" position is reached.)				
o. *Brake action:* Abrupt or jerking stops. (Check at various positions in hoistway in both up and down directions of car.)				
p. *Acceleration and retardation:* Excessive or not smooth.				
q. *Car enclosure:* Not securely fastened to platform; bolts, screw, or other fastening loose or missing; opening in grille or large mesh not properly protected; part of enclosure can be deflected by passenger or freight to cause car to strike counterweight or obstruction in hoistway; capacity plates, inspection certificates, operating and emergency instructions not posted in car; unauthorized signs or posters on walls; deteriorated or damaged wall or ceiling panel; unauthorized alterations or additions to car that have changed its weight.				
r. *Car lighting fixtures:* Insecurely fastened or otherwise defective; provide insufficient illumination.				
s. *Car fans:* Insecurely fastened, inoperative, provide insufficient ventilation.				

E	U	R	D

t. Operator seats: Insecurely fastened, ineffective or damaged brackets, cracked or splintered seats.

u. Car emergency exits: Not properly closed, emergency key for side exit not kept on premises in location readily available to authorized persons; emergency key available to general public; top exit not openable from both inside and top of car without use of tool; side exit openable from inside car by means other than emergency key, or from outside car by means other than nonremovable handle; obstruction preventing opening exits from either side; electric contact for side exit does not prevent car operation with exit open. (Open both top and side exits and determine if any in improper condition. With side exit open move car operating device to "On" position. The car shall remain stationary.)

v. Floor signal devices: Do not function properly to indicate floors at which stops are to be made, or to signal car approach to a landing at which a landing signal registering device has been actuated.

w. Operating hand ropes: Rope contains excessive number of broken wires, or is weakened from wear or other defects; rope locks are inoperative; rope does not actuate centering device properly to stop oar (freight elevator) or rope does not remain in neutral position when centered.

2. *INSPECTION OUTSIDE OF HOISTWAY:*

a. Hoistway enclosure keys and landing buttons: Hoistway enclosure, door panel, or grating loose, bent, or corroded; grille or large-mesh grating not covered with screen of 1/2-inch mesh or less; emergency hoistway door key not in its proper location, or

E	U	R	D

glass cover to key receptacle broken; service key not kept in suitable location on premises and available only to authorized personnel, or will open hoistway door other than at the service (parking) landing, or will open the service landing door when car is not within the landing zone; one or more landing buttons stick; automatic operation pushbutton, when actuated, does not call car to landing; signal button for car switch operation does not function. (Check at each landing as appropriate.)

b. *Counterweight enclosure and wire rope:* Enclosure access door not self-closing, can be opened from outside by means other than with proper key, cannot easily be opened from inside without use of key or other instrument; the door or its locking mechanism is otherwise defective; wire rope shows excessive wear, broken wires, or other defects; rope is inadequately lubricated.

3. *INSPECTION FROM TOP OF CAR:*

a. *Counterweight fastenings and buffer:* Cotter pins or lock nuts not properly in place at both top and bottom of counterweight through-rods; counterweight frame bent or otherwise damaged; oil in counterweight oil buffer not at proper level; buffer holding-bolt loose or missing; for drum machine, car counterweight is not properly above machine counterweight with a clearance of at least 8 inches; rim or arm of counterweight sheave cracked; rim of sheave chipped or excessively worn.

b. *Wire rope fastenings and equalizers:* Rope fastenings on counterweight, car crosshead, or governor and safety rope show; one or more wires broken or bulge out and carry no load in a pitch length or lay of rope nearest shackle; corrosion in a pitch length or lay of rope nearest shackle; wires damaged from overheating, acidulous action of cleaning material, or other causes where the rope enters the socket, evidence of rope pulling in socket; shackle deformed or critically weakened from wear. Car-end shackles for hoisting ropes of power elevator having driving machine with one-to-one roping have not been reattached or replaced within 24 months for machine located over hoistway, or within 48 months for machine located below or at side of hoistway. Rope equalizer not properly at center of its range of motion when car is at its center of travel; equalizer will reach limit of its range of motion in either direction when car is at limit of travel.

c. *Wire rope:* Rope for car, counterweight, governor, or safety device not of proper diameter, material, or construction (refer to rope tags, crosshead data plate, or manufacturer's specification); rope has lost its "lay," or has been pinched, crushed, or otherwise mechanically damaged, or shows excessive wear or broken wires. Rope for car, counterweight, or safety device is inadequately lubricated; governor rope shows evidence of having been lubricated. (Governor ropes shall not be lubricated.)

d. *Normal slowdown and stop switches:* Switch and engaging can be misaligned; cam not properly ridged; holding bolt for switch or cam loose or missing; normal slowdown switch does not function properly before normal stop switch is actuated; normal stop switch does not function properly before final-limit switch is engaged;

E	U	R	D

E	U	R	D

where a floor selector in the machine room is used as a normal stopping device, this device does not function properly.

e. *Upper final-limit switch:* Switch and engaging cam misaligned; cam not properly rigid; holding bolt for switch or cam loose or missing; fastening for conduit, junction box, outlet box, or cover for junction or outlet box loose or missing; car can be operated by normal operating device with upper final-limit switch opened by hand; function of final-limit switch interferes with normal stop, or final-limit switch does not open with car as close to upper terminal landing as possible. (To determine: Position car level with upper terminal landing and measure distance from end of cam to center of limit roller. Final limit switch shall open or as close to upper terminal landing as possible without interference with normal stop.

f. *Car and counterweight horizontal clearance: power elevators:* Clearance less than: 3/4 inch between car and hoistway enclosure, except between car and landing threshold where steel guides are used; 1/2 inch between car and landing threshold where steel guides are used; 1 inch between car and counterweight; 3/4 inch between counterweight and counterweight screen or hoistway enclosure; 2 inches between cars in adjacent hoistways. (Take actual measurements when safety permits, otherwise determine clearance by close observation.)

g. *Guide rails (car and underweight):* Misaligned joint, bend, or deviation; lubricant or coating insufficient, excessive, or of improper type; lubricant on rails that should

E	U	R	D

be free of lubricant (when roller guides are used, the rails shall be dry); rail insecurely fastened to bracket or bracket insecurely anchored; rail surface worn, rough, or scored (if rails have been scored by repeated application of safeties, determine cause. This may be due to improperly adjusted crosshead spring, improper controller adjustment, improperly adjusted or worn brakes, or weak releasing carrier spring).

h. *Door operating devices:* Motor, engine, cam, switch, chain, sprocket, or other operating mechanism excessively worn, insecurely fastened, inadequately lubricated, or otherwise defective; counterweight for door or gate not properly boxed; reting cam for car not in proper alignment with interlock switch arm, or its travel is insufficient to ensure operation of lock; semiautomatic door or gate for hoistway opening of freight elevator does not close properly as car leaves landing.

i. *Governor-rope shackle releasing carrier:* Parts rusted or caked with dirt; broken spring; improper tension; safety rope does not properly fill grooves of safety drum, or is improperly wound.

j. *Special slowdown or car-latching device (top of hoistway):* Loose, broken, excessively worn. (If possible and safe, operate by hand.)

4. *OVERHEAD INSPECTION:*

a. *Wire rope:* Excessive wear or broken wires, inadequate lubrication, or other defect, rope anchorage loose, deformed, or excessively corroded.

b. *Overhead sheaves:* Chipped or excessively worn rim; cracked rim or arm; eccentricity

E	U	R	D

of motion; excessively worn or inadequately lubricated bearing.

c. *Overhead supports:* Unsound conditions, dislodged, fastenings insecure, loose, missing, or excessively corroded.

d. *Overhead grating, screen, or platform* (CAUTION: Before stepping on any overhead grating, screen, or platform, examine supports and fastenings carefully to see that it is rigid and strong enough to safely hold your weight): Loose, broken, rusted, rotted, or otherwise deteriorated members); loose or missing fastenings; openings greater than 3/4 inch between bars of metal grating, or greater than 1 inch diameter in perforated sheet metal or fabricated openwork.

e. *Governor:* Rope does not run clear of governor jaws during normal operation; not ample room for rotation of balls through complete range of possible motion; eccentricity of motion of governor sheave; rope slides in sheave when car is stopped or started; excessively worn part, inadequate lubrication, or moving part not operating freely; rope grip jaws do not engage rope properly when governor is tripped by hand; field-regulating switch and stopping contact in control circuit do not operate at or before time governor rope grip jaws drop; potential switch does not open at or before time governor rope grip jaws drop; governor seal is missing or broken.

5. *INSPECTION IN MACHINE ROOM:*

   a. *Machine room:* Tripping hazard; excessive oil, grease, or dirt on machinery or floor; equipment or material not essential to maintenance stored in machine room; evidence of water leak from roof or other source.

E	U	R	D

b. *Electric hoisting machine:* Rim or arm of traction sheave or winding drum cracked or chipped, or groove excessively worn; all ropes do not seat to same depth in grooves; bearing excessively worn or inadequately lubricated; traction sheave shows loss of traction or is misaligned; for winding drum there is not at least one full turn of rope on drum when car and counterweight are at their respective limits of over-travel.

c. *Electric hoisting machine gears:* Gear or worm excessively worn, or rubbing surface pitted or scored; gear guard, where required, is insecure and not intact.

d. *Electric hoisting machine gear backlash and lubrication:* Excessive thrust in worm shaft or backlash in worm gear; oil in gear case not at proper level, contains metallic particles, does not have sufficient body, is gummy or semisolid, or is rancid.

e. *Electric hoisting machine bearings:* Evidence of lost motion, excessive wear, or damaged rollers or balls; oil reservoir not properly filled with clean oil: oil ring, chain, or other method of feeding not operating freely.

f. *Electric hoisting machine limits (winding drum machine)*: Position of car after being brought to rest by normal machine-limit stop shows evidence it will travel far enough beyond upper or lower landing to open final machine limit or final hoistway limit; machine final limit or hoistway final limit interferes with machine normal stop or does not open with car as close to terminal landing as possible.

g. *Electric hoisting machine motor and generator:* (CAUTION: Before touching electric current carrying parts, be sure main

E	U	R	D

switch is open): Excessive end-play shaft; evidence of bearing overheating, excessively worn, broken ball or roller, or insufficient or excessive lubrication; commutator or slip ring burned, pitted, grooved, scored, or oily; commutator has high mica; brush holder, brush, or spring improperly adjusted, or has loose connection; brush excessively worn, chatters, or does not bear uniformly; windings oily or excessively dirty; grease or oil deposits inside housing; terminal wiring has cracked or broken insulating material or insecure fastenings.

h. *Electric hoisting machine brake:* Insufficient or excessive spring pressure; lining or operating mechanism excessively worn; drum scored; shoe drags on drum with car running, or clearance between shoe and drum is greater than necessary to secure free running; chattering; improper dynamic braking; motor windings oily or excessively dirty. (Check for harsh or abrupt brake action when car is making several stops in both hoisting and lowering directions.)

i. *Controller mechanism:* (CAUTION: Before touching electric current carrying parts, be sure main switch is open): Relays and contactors, contact burned, pitted, or excessively worn; gummy oil deposits on laminated core surface; sticks in closed position; corroded mechanism; residual washer broken or flattened. Accelerating switches do not close in proper order. Reverse phase relay inoperative. (To check relay operation, remove and replace all main line fuses connected in the circuit one at a time.)

j. *Controller wiring fuses:* Loose or broken wiring connection; cracked or warped re-

E	U	R	D

sistance grid or tube; lint, dust, or other debris, particularly on resistance grids; improper-size fuse; fuse clips bridged, or fuse rendered useless by wire of metal strip connected ferrule to ferrule; fuse in neutral or grounded conductor. (There should be no fuse in neutral or grounded conductor.)

k. *Driving belts:* Belt badly worn or burned, or shows appreciable breaks in surface; belt shifting forks excessively worn or do not shift belt fully to proper pulley.

l. *Hydraulic machine sheaves:* Chipped or excessively worn rim; cracked rim or arm; excessively worn or inadequately lubricated bearing; eccentricity of motion; guide rail or guide shoe of traveling sheaves worn or loose to the extent of permitting appreciable play of parts, or is inadequately lubricated.

m. *Hydraulic machine piston rods:* Exposed portions show excessive wear, corrosion, or misalignment; equalizing washer on equalizing cross-head not properly in place under piston rod nut; loose, broken, or missing fastening for plunger of plunger elevator; other deficiencies that can be determined without dismantling unit or part.

n. *Hydraulic machine cylinders:* Appreciable vibration or movement of cylinder support when machine is operating; corrosion or small cracks, particularly at bolt holes on flanges; appreciable leakage from piston rod or plunger packing when machine is under pressure; air-bleeder cock clogged or otherwise defective.

o. *Hydraulic machine valves (terminal and operating valves):* Appreciable leakage, bent stem, loose or missing handle; counterweight on shut-off valve insecurely fas-

E	U	R	D

tened to arm; automatic cut-off, rope, or rope guard show defects or improper operation. (Piston or valve leakage indicated if car creeps when operating device is in OFF position.)

p. *Hydraulic machine pump:* Loose packing or appreciable leaking of pump or valve; relief valve, automatic bypass valve, or safety valve in improper operating condition. (Make safety valve operate by raising pressure slightly; 10% above normal working pressure should be enough. Be sure pressure is returned to normal after inspection.)

q. *Hydraulic machine discharge tanks:* Leaks or corrosion; vent plugged; cover not properly in place on open type.

r. *Hydraulic machine pressure tanks:* Leaks, corrosion, or incipient cracks; pressure gauge damaged, difficult to read, or registering incorrectly; defective or improperly operating vacuum valve; loose, broken, or plugged water gauge.

s. *Handrope fastenings:* Rope fastenings, sheave, sheaveguard, shaft, or fastening on operating shaft show loose, broken, excessively worn, or missing part; equipment improperly grounded.

t. *Slack-rope devices located under the drum (winding drum machine only):* Bent, loose, broken, or excessively worn part, or improper operation; detector bar not set close enough to drum to cause device to function if car is obstructed in its descent; improper operation. (Check operation by tripping device by hand with machine running. On a rope-driven machine, the device shall cause the electric power to be disconnected from the driving machine motor. On a belt-

E	U	R	D

driven machine, the device shall shift the driving belt to the idler pulley.)

6. *INSPECTION IN PIT:*
   a. *Car and counterweight guide shoes and safety jaws:* Insecure, misaligned, or maladjusted guide shoe or fastening; safety jaw, block, roller; wedge, or guide shoe sufficiently worn to interfere with proper application of safety. (Safety jaws shall not touch guide rails under normal operating conditions.)
   b. *Car and counterweight safeties—type B—hand pull-out:* Drum mechanism excessively worn, corroded, or inadequately lubricated; drum and shaft cannot be rotated by manually pulling on drum rope to bring safety jaws into contact with guide rails; drum rope shows corrosion, flattened strand, inadequate lubrication, excessive wear or number of broken wires, or is different from that specified by manufacturer; pull-off of rope materially exceeds that for which safety was designed, or there are insufficient turns remaining on drum after hand pull-out to prevent danger of rope being pulled from drum when safety is applied by action of governor. (Reset safety after completing inspection. Rope shall be closely and uniformly wound on drum without remaining slack.)
   c. *Car and counterweight oil buffers:* Holding bolt loose or missing; low oil level; gage does not indicate oil level properly; pet cock plugged; plunger does not return properly after being depressed; piston shows excessive side play; non-return switch does not open when buffer is compressed more than approximately 3 inches.
   d. *Car and counterweight spring buffers:* Vertical misalignment; deformed, or has per-

E	U	R	D

ceptible set; improperly seated in cup or mounting; support interferes with compression of buffer; striker plate damaged or displaced.

e. *Counterweight guards:* Loose, improperly placed, bent, or otherwise damaged to extent that they may interfere with movement of car or counterweight.

f. *Car sling:* Loose or missing nut or bolt; frame distorted; platform not approximately level.

g. *Governor-rope tension sheave:* Sheave, shaft, or bearing excessively worn, damaged, or inadequately lubricated; eccentricity of sheave; frame does not slide properly in guide; insufficient clearance between bottom of frame and pit floor.

h. *Compensating cable tension sheave:* Sheave, shaft, or bearing excessively worn, damaged, or inadequately lubricated; eccentricity or misalignment of sheave, insufficient clearance between bottom of sheave or sheave frame and pit floor; safety switch inoperative. (Try safety switch when car is in motion. Opening this switch shall stop car.)

i. *Compensating cable tension-sheave locking device:* Loose, worn, broken, corroded, or other defect which may prevent its functioning properly.

j. *Lower final-limit switch:* Switch and engaging cam misaligned; cam not properly rigid; holding bolt for switch, cam, conduit, or junction box cover loose or missing; improper operation (car should not be openable by normal operating device when lower final-limit switch is open; final-limit switch should not interfere with normal stop; final-limit switch should open with

E	U	R	D

car as close to lower terminal landing as possible).

l. Pit floor: Refuse or water, material stored in pit, pit light burned out.

m. Special speed-control devices (at bottom of hoistway): Loose, broken, excessively worn, or do not function properly. If possible, operate by hand.

n. Traveling cables: Rests on pit floor with car at lower limit of travel; outer coating shows abrasion or protruding broken wires; loose, worn or otherwise defective anchorage at car or hoistway end; anchorage or cable will contact wall, beam, or landing sill.

7. *INTERNAL INSPECTION:*
Piston rods—hydraulic elevator: Piston rod, stop block used to limit piston travel, plunger shoe or bypass corroded, cracked, broken, excessively worn, otherwise defective, or does not function properly; holding bolt for plunger head loose or missing. (Inspection made after piston rods exposed thoroughly cleaned.)

8. *TESTS:*

a. Governor-operated car and counterweight safeties—NO-LOAD test at slow speed: Test made with no load in car and car (or counterweight) traveling at slow speed in the down direction. Select a position for car to start test such that when car (or counterweight) comes to rest by action of the safety, safety will be in reach for close visual inspection from pit floor or from a temporary platform or ladder placed in pit. Have operator move car at slowest operating speed in down direction (up direction when testing counterweight safety). While car is in motion, manually trip governor, causing safety to engage guide rails and

E	U	R	D

stop car. Before releasing safety inspection mechanism, inspect for the following deficiencies: all parts of elevator equipment subject to damage for broken or inoperative parts; improper functioning of governor; damage to governor rope; improper condition of rope releasing carrier and other linkages; defects in safety drum mechanisms; insufficient turns of rope remaining on drum for wedge-clamp or gradual wedge-clamp safeties; all hoisting ropes not in their respective grooves of sheaves or drums; brake not applied; condition of guide rails that may prevent proper operation of safeties; slack hoisting ropes on winding drum machines. Release and relatch governor rope jaws; release safety; rewind safety rope on drum, keeping sufficient tension on drum or governor rope to ensure smooth even winding, until releasing-carrier shackle is in its proper position in carrier and there is not excessive slack in the safety-drum rope; move car a few inches and remove grit, piled up by lower edge of safety jaw, from rail. Report if safety made deep marks in rail or badly marred rail.

b. *Broken-rope (instantaneous) operated car and counterweight safeties—NO-LOAD test:* Test made with no load in car and car (or counterweight) at a position near bottom of hoistway or at the upper limit of travel. Test of car safety may be made by either of the following methods. Counterweight safety may be tested in a similar manner except that car is moving in up direction and counterweight is moving in down direction when counterweight safety is applied.

E	U	R	D

With car near the bottom of hoistway: Place a support under car and lower car onto support until there is sufficient slack in hoisting ropes so that when support is removed the safety can act to stop car before all slack in hoisting rope is taken up. Place wood blocking in pit to a height sufficient to stop car and avoid injury to hoisting ropes should the safety fail. Remove support under the car by means of a rope or block and fall. As car drops, the safety should grip the rails and stop car within a short distance.

With car at the upper limit of travel: by means of a chain block or fall.

Before releasing safety mechanism inspect for the following deficiencies: all parts of elevator equipment subject to damage for broken or inoperative parts; improper functioning of safety mechanism and related equipment; all hoisting ropes not in their sling from overhead beams raise car until there is sufficient slack in hoisting rope so that when car is lowered by the chain fall, safety can act to stop car before all slack in hoisting ropes is taken up. When necessary, place blocking under counterweight to obtain sufficient slack in hoisting ropes. Then lower car rapidly by means of the chain fall. As car drops, safety should grip rails and stop car within a short distance.

Before releasing safety mechanism, inspect for the following deficiencies: all parts of elevator equipment subject to damage for broken or inoperative parts; improper functioning of safety mechanism and related equipment; all hoisting ropes not in their respective groves of sheaves or

E	U	R	D

drums; brake not applied; condition of guide rails that may prevent proper operation of safeties.

Release safety and relatch tripping device. Move car a few inches and remove grit, piled up by the roll or wedge of the safety, from rail. Report if safety made deep marks in rail or badly marred rail.

c. *Slack-rope device on top of car—operation test:* Test made with no load in car. Obtain enough slack in hoisting rope to cause slack-rope device to function. This may be done by lowering car onto a suitable support or supports placed in pit until tension in hoisting ropes is decreased to the point at which the slack-cable switch will operate. Operation of the device should automatically remove power from the elevator driving machine motor and brake.

d. *Governor-operated counterweight safety—NO-LOAD test at rated speed:* Test made in manner as specified above except that it shall be made with the counterweight traveling at rated speed in the down direction as counterweight safety is applied. The stopping distance or slide of the counterweight safety shall be recorded.

e. *Governor—overspeed test:* Remove governor rope from governor sheave and drive governor by a hand or motor driven device with gradual acceleration until a speed is reached at which governor trips. Measure governor tripping speed by means of a tachometer and check governor tripping speed arc setting of the governor overspeed switch for conformity with the following requirements.

TRIPPING SPEEDS FOR CAR SPEED GOVERNORS: Governor shall be set to trip for overspeed at not less than 115% of rated speed.

E	U	R	D

TRIPPING SPEEDS FOR COUNTERWEIGHT SPEED GOVERNORS: Governors shall be set to trip at overspeed greater than, but not more than 10% above, that at which a speed governor is set to trip.

SETTING OF CAR SPEED-GOVERNOR OVERSPEED SWITCHES: The setting of the governor overspeed switch shall conform to the to following. For car rated speeds more than 150 fpm, up to and including 500 fpm, the overspeed switch shall open in the down direction of the elevator at not more than 90% of the speed at which the governor is set to trip in the down direction; for car rated speeds more than 500 fpm, the overspeed switch shall open in the down direction of the elevator at not more than 95% of the speed at which the governor is set to trip in the down direction; the switch, when set in either condition above, shall open in the up direction at not more than 100% of the speed at which the governor is set to trip in the down direction. EXCEPTION: The speed-governor overspeed switch may be set to open in the down direction of the elevator at not more than 100% of the speed at which the governor is set to trip in the down direction subject to the following: a speed-reducing switch of the manually reset type is provided on the governor which will reduce the speed of the elevator in case of overspeed and which shall be set to open as specified above for car rated speeds more than 150 fpm, up to and including 500 fpm, and for car rated speeds more than 500 fpm; subsequent to the first stop of the car following the opening of the speed-reducing switch, the car shall remain inoperative until the switch is manually reset.

Total Items: 8 Raw Total %

E	U	R	D

Findings (comment on each checkpoint and summarize to justify scoring):

____

____

____

____

____

____

Section 9. Food Preparation and Service Equipment

SCOPE: Kitchen equipment used for cooking, boiling, steaming, toasting, and other warming, chilling, and serving of hot or cold foods.

1. *USER'S COMMENTS:* Ask operator or supervisor for comments on equipment performance before starting assessment.
2. *SANITATION:* To safeguard personnel from food contamination and food-borne diseases, it is important that safe sanitary conditions be restored after each assessment.
3. *DEGREASING:* Report danger areas where grease deposits might cause fire, food contamination, or obstructions to ventilation, drainage, or liquid flow.
4. *OPERATING CONTROLS:* Improper operation through complete cycle, improper "On" and "Off" operation.
5. *FIRE PROTECTIVE DEVICES:* Incorrect temperature fusible links, improper setting

E	U	R	D

of thermal unit or releasing device, excessive high temperature of fan stop device.

6. *THERMAL INSULATION AND PROTECTIVE COVERINGS:* Open seams, breaks, missing sections, missing or loose fastenings.
7. *BURNER ASSEMBLIES:* Loose, damaged, or missing connections and parts, leakage improper fuel–air mixtures, incorrect height and position of pilot light, improper baffle adjustment, causing impingement, dirty heat transfer surfaces, nonuniform flame spread, misalignment, clogged jets, orifices, and valves, dirty oil filters, defective oil wicks, oil rings, and pots. Test heating output at each step of multi-step heating level; tighten loose connections; repair or replace damaged or missing parts; adjust fuel air mixture to produce blue flame; adjust pilot light; replace black or burnt-out pilot lamps; reset baffles; clean heat transfer surfaces and ignition devices; correct misalignment; clean openings in jets, orifices, and valves; replace dirty oil filters, defective oil burner, other parts of assemblies.
8. *COMBUSTION CHAMBERS:* Deformations, breaks, cracks, wear, water and flue gas leakage, burnt-out grates, defective coal-feed mechanisms, broken latches and hinges, door misalignment and poor fit, soot deposits, clinkers, ashes.
9. *ELECTRICAL HEATING UNITS:* Burned, pitted, or dirty electrical contacts, short-circuited sections of elements, low voltage in electrical circuits, dirty reflective heat transfer surfaces. Test heating output at each step of multiple-step heating levels, clean contacts and heat transfer surfaces, tighten loose connections.
10. *METERS, PRESSURE GAUGES, INDICATING AND RECORDING INSTRUMENTS,*

E	U	R	D

THERMOMETERS AND THERMOSTATS: Leaks, cracked dial-cover glasses, defective gaskets, moisture behind glasses, mechanical damage.

11. *DUCTS, SMOKEPIPES, HOODS, DAMPERS, AND DRAFT DIVERTERS:* Clogging, soot, dirt, grease, other deposits; loose connections and parts.
12. *DRAIN, WATER METER RELIEF, SUPPLY, STEAM, BYPASS, AND SAFETY VALVES:* Clogging, scale, loose connections, leakage, open bypass valves, defective operation. Remove clogging; tighten connections; close bypass valves; immediately report defective operation of safety and relief valves.
13. *STEAM COILS:* Inadequate steam pressure, clogging, scale leaking, loose connections, misalignment, water hammer, airbinding, nonuniform heat spread, below-normal temperature readings. Remove scale and dust; tighten connections; relieve water hammer and air-binding.
14. *STEAM AND WATER PIPING AND PUMPS:* Clogging, leaks, loose connections and fittings, deformed and unsafe parts, excessive noise and vibration, other deficiencies Verify water temperatures using mercury thermometer.
15. *CONTAINERS OF FOOD AND DRINK MAKERS, STEAM, BAKING, FRYING, COOKING, HEATING, AND TOASTING DEVICES:* Poor appearance, dirty, grease and other deposits, leaking, defective gaskets, cracked or chipped glass containers, loose connections, defective draw-off faucets, filters, stoppers, broken handles, improper level, poor fit of doors and covers, missing or damaged parts, improper heating, defective operation.

E	U	R	D

16. *MIXERS AND JUICERS* (partially disassemble to examine internal parts): Dirty; obstructions; misalignment; excessive vibration; broken, loose, or missing parts: loose belts or other deficiencies.
17. *DISPLAY FIXTURES:* Poor appearance, dirty, to painted surfaces, defective gaskets, poor fit of doors, damaged hardware, cracked or chipped glass, loose, broken, or missing parts, defective, lamps at improper level, clogged or leaking trays, drains, or coils, defective operation.
18. *PEELERS, SLICERS, CHOPPERS, AND CUTTERS:* Misalignment, excessive dirty, noise and vibration, inaccuracy of thickness adjustors, inadequate protective guards, worn, dull, or other damage to abrasive disks and cylinders, loose, missing, or broken parts of cutting edges.
19. *BUTCHER BLOCKS AND TABLES:* Extreme surface roughness, splinters, deep or numerous crevices and cracks, stains, odors, and evidence of possible harboring of bacteria, other components for loose joints and parts, misalignment.
20. *PORTABLE FOOD WARMERS:* Carts and miscellaneous kitchen equipment: dirty, need of degreasing and sanitizing, loose bolts, missing or damaged parts, difficult movement, deteriorated paint, other deficiencies.

Total Items: 20 Raw Total %

Findings (comment on each checkpoint and summarize to justify scoring):

____

____

E	U	R	D

Section 10. Heating Equipment

SCOPE: Heater and controls console, unit heater (gas and oil fueled), unit heater (steam and hot water), heat pumps.

1. *HEATER AND CONTROLS CONSOLE:* Check the following: air line lubricator; lines to and from unit for steam, water, or oil leaks; dust filter for dust accumulation; alignment of pulleys, check condition of belt; electric motor bearing for lubrication; steam valves for proper operation; steam trap, check motor bearings for noise or wear; buss bolts for tightness, also inspect wiring and electrical controls for loose connections, charred, broken, or wet insulation; evidence of short circuiting and other deficiencies.
2. *UNIT HEATER (GAS AND OIL FUELED):* Check the following: heater deflector fins and element; fan and lubrication, burner, chamber, thermocouple and control; pilot or electric ignition device; also inspect vent and damper operation, operate unit and check burner; op-

E	U	R	D

eration of safety pilot, gas shut-off valve, and other burner safety devices.

3. *UNIT HEATER (STEAM AND HOT WATER)*: Check strainer ahead of valve, valve head, and seats for wear and cutting. Steam quality should be examined for foreign matter if valves are being damaged. Check steam gauges, check non-safety or pressure relief valve for relieving and seating, diaphragms for failure, binding of valve stem, heater deflector fins and element, fan operation and lubrication, check weighted lever or spring control tension.
4. *HEAT PUMPS:* Inspect piping for evidence of leaks and vibration; inspect all wiring for deterioration, and tighten electrical contacts; check for corrosion, mounting bolts and tighten if needed; check crankcase heater, fan operation for vibration or excessive noise, and motor for lubrication; check refrigerant levels, and repair leaks if loss of refrigerant is detected. Check temperature drop across condensing coil; check air intake and screens and coil surfaces; check that reversing valve is energized in the "heat" mode and deenergized in the "cool" mode; check oil.

Total Items: 4 Raw Total %

Findings (comment on each checkpoint and summarize to justify scoring):

____

____

____

Section 11. Plumbing Systems

E	U	R	D

SCOPE: Plumbing and piping equipment and systems.

1. Check for possibility of contamination by connection of potable water to waste systems, improper gap between potable water fixtures and waste pipe rim of sink, or similar fixture.
2. Check piping for rust. corrosion, clogging, and damage.
3. Check valves, fixtures for leakage, proper operation, corrosion, scale, clogging, and damage.
4. Check insulation for open seams, damage, missing sections. and improper or missing fastenings.
5. Check platforms, pedestals, and supports for corrosion, missing or loose connections, defective parts, alignment and damage.
6. Check: sanitary and drain piping for accumulations in strainers, proper drainage, open vents, adequate water seal in traps, and siphonage of traps and leaks.
7. Check floor drains clogging and backwater valves for clogging, leaking connections, damage, and presence of sewer gas and odors.
8. Check grease and lint traps for grease, fibers, hair, other obstructions, leaks, and damage.
9. Check piping identification for legibility and correctness.
10. Inspect all exposed piping for leakage, corrosion, loose connections, and damage.
11. Inspect underground piping for leakage, ponding erosion, and settlement.
12. Check buried valves for bent stem, leakage, corrosion, and proper operation.
13. Check exposed valves for bent stem, leakage, corrosion, and proper operation.

E	U	R	D

14. Inspect water meters for accuracy, leakage, corrosion, broken glass, moisture behind glass, settling, and proper operation.
15. Inspect hydrants and hydrant shut-off valves (SOV) for missing caps, broken or missing chains, damaged threads, missing or damaged guards and identification markings.
16. Check hydrants for rust and corrosion.
17. Check valve and meter pit manholes and roadway boxes for rust, corrosion, and damage.
18. Check manhole frames, covers, and ladder rungs for rust, corrosion, loose or missing rungs, and other damage or deficiencies.
19. Check concrete and mortar joints in manholes.
20. Check overflow pipes for water entrance, rust, and damage to screen.
21. Check walkways, guardrails, stairs, and ladders for rust, corrosion, broken or missing parts.

Total Items: 21 Raw Total %

Findings (comment on each checkpoint and summarize to justify scoring):

____

____

____

____

____

____

E	U	R	D

Section 12. Ventilating and Exhaust Air Systems

1. *VENTILATING AND EXHAUST AIR SYSTEMS:*
   - *a.* Check lubrication of fan shaft bearing and/or electric motors as applicable.
   - *b.* Check belt condition.
   - *c.* Inspect wiring and electrical controls for loose connections, charred, broken, or wet insulation, short circuits, etc.
   - *d.* Check motor for excessive heat and vibration.
   - *e.* Inspect for rust and corrosion.
   - *f.* Check screen and vent cleanliness on roof, as applicable.

E	U	R	D

g. Check ducts, collectors, smokepipes, and hoods for clogging, soot, dirt, and grease.
h. Inspect cover guards, supports, covers, etc.

2. *HEATING AND VENTILATING UNITS:*
   a. Check air filter.
   b. Check heater operation.
   c. Check lubrication of shaft and motor bearings.
   d. Check wiring and electrical controls for loose connections, charred, frayed, or broken insulation.
   e. Check for rust and corrosion.
   f. *Air Handler Unit:* check fan blades for dust buildup and cleanliness; fan blades and moving parts for excessive wear; fan RPM to design specifications; bearing collar set screws on fan shaft to ensure they are tight; dampers for dirt accumulations, felt, damper motors, and linkage for proper operation; mechanical connections of dampers; lubrication coils for cleanliness. Check coils for leaking, tightness of fittings; condensate pans and drains for cleanliness; belts for wear; rigid couplings for alignment on direct drives and for tightness of assembly; flexible couplings for alignment and wear; freeze-stat for proper operation; interior of unit for cleanliness.

Total Items: 2 Raw Total %

Findings (comment on each checkpoint and summarize to justify scoring):

____

____

____

E	U	R	D

Section 13. Hot Water Systems

SCOPE: Water heater—electric, water heater—steam, water heater.

1. *HOT WATER HEATER—ELECTRIC:* Check tank for sediment, manually check operation of safety valve, check for corrosion, check all connections—electric and water, check operation and setting of aquastat, check hot water temperature, check amperage draw of upper and lower elements and compare to name plate data, check element contacts for proper closing under load.
2. *HOT WATER HEATER—STEAM:* Inspect element heater and exterior tank, including fittings, manholes, and handholes for leaks and signs of corrosion. Test pressure relief valve, remove tank inspection plate and inspect condition of interior of tank: record the size and depth of pitting, presence of cracks and condition of openings, fittings, welds, rivets and joints. Check condition of heat exchanger element, inspect condition of epoxy tank lining, check all gaskets, manhole inspection plates, and bolts. Check tank for leaks, check condition of traps and strainers. Check pump, con-

E	U	R	D

trols, switches, and starters. Check condition of pump seal or packing, inspect sight glasses, valves, fittings, and drains, inspect structural supports for damaged insulation or covering, observe temperature control operation. Adjust controls as required, as tank is returned to service. (Hydrostatic testing is not included in the purview of this checkpoint.)

3. *WATER HEATER:*
   a. Check operation of automatic controls through complete cycle.
   b. Inspect thermal insulation and protective covering.
   c. Check burner assembly for loose, damaged, or missing parts and proper fuel–air mixtures as applicable.
   d. Check electrical heating element and controls as applicable.
   e. Inspect wiring and electrical controls for loose connections, charred, broken, or wet insulation, short circuits and other deficiencies.
   f. Check safety valve for rust, corrosion, and proper operation.
   g. Check for leaking valves, steam traps, and other defects, as applicable.
   h. Inspect for rust and corrosion.

Total Items: 3 Raw Total %

Findings (comment or each checkpoint and summarize to justify scoring):

____

____

____

E	U	R	D

____

____

____

____

____

____

____

____

____

____

____

____

____

____

____

Section 14. Electrical Systems

SCOPE: Electrical light and power systems for buildings, 600 volts and below, exclusive of supply and utilization equipment. Building electrical systems extend from the main service switch or breaker to points of connection for lights, motors, or other utilization. Included are wire and cable runs; busways, raceways, and wireways; tubing

E	U	R	D

and conduit; switchboards, panelboards, and cabinets; circuit protective and switching equipment; junction and cutout boxes; insulators, supports, and wiring devices; and other items not associated with supply or utilization equipment, such as metering, motor controls, and starters.

1. *CABLE AND WIRING:*
   a. Dirty, poor ventilation, detrimental ambient conditions, presence of moisture, greases, oil, chemical fumes.
   b. Improper or unauthorized connections and dangerous temporary connections.
   c. Damaged wiring devices, defective insulators, cleats, and cable supports; broken or missing parts or exposed live parts.
   d. Excessive cable sag and vibration, crowded cable spacing, excessive number of conductors in conduit and raceways.
   e. Evidence of overheating, grounds, and short circuits; overheated splices, damaged or defective insulation.
   f. Need for painting of non-current-carrying parts subject to corrosion.
   g. Unsafe, unreliable cable and wire to lighting and power panels.
   h. Determine if there are fuses, switches, and other sources of discontinuity in the neutral wire of grounded AC systems.
2. *PANELBOARDS:*
   a. Dirty, corroded, signs of overheating, and need for touchup painting. Unposted or illegible instruction, identification charts, circuit diagrams, and feeder schedules.
   b. Loose or inadequate connections.
   c. Switches require lubrication; operate improperly. Knife switches and fuse chips improperly aligned.
   d. Check fuse ratings and ground connections.

E	U	R	D

e. Determine if there are fuse jumpers and dangerous temporary connections.

3. *SWITCHES AND BREAKERS:*
   a. Poor condition of contacts, contact misalignment, signs of overheating.
   b. Defective operation (try manually and electrically when practicable).
   c. Spot check and report voltage on branch circuits, feeders, and convenience outlets serving motors, heaters, and other utilization equipment.
   d. Measure ground resistance at various points on metal conduit and grounding systems.
   e. Measure insulation resistance in feeders and branch circuits to terminals of utilization equipment between each conductor and ground. Insulation resistance should not be less than 300,000 ohms.

Total Items: 3 Raw Total %

Findings (comment on each checkpoint and summarize to justify scoring):

____

____

____

____

____

____

____

E	U	R	D

Section 15. Lighting

SCOPE: General and supplementary lighting in building interiors. Includes fixtures, lamps, switches, outlets, cords, and other equipment associated with lighting only, generally installed between lighting panels and light-producing equipment. Branch circuit or feeder wiring and lighting panels are not covered.

1. *LIGHTING FIXTURES:*
   - *a.* Inadequately supported, insecure, and improperly located, evidence of unauthorized removal and relocation
   - *b.* Incorrect types installed in hazardous locations; change in facility use requires replacement.
   - *c.* Improperly located in closets. (Should be above door or in ceiling, and not serviced with cord pendants.)
   - *d.* Cracked or broken luminaries and fixture parts, missing pullcords, metal pullchains not provided with insulating links.
   - *e.* Indications of objects being supported from, hung on, or stored in fixtures.
   - *f.* Evidence of overheating, undersized, or other damage to sockets, exposed or damaged connecting wiring.

E	U	R	D

   g. Check exposed live wires, undersized or overheated sockets, broken or missing pull-cords, cracked glass luminaries, and missing fixture parts.

2. *LAMPS:*
   a. Oversized, blistering, loose base. Thermal cracks from contact with fixture, bare lamps in hazardous locations. Poor socket types for special application lamp connections. Improper types for special applications.
   b. Operation of fluorescent fixtures shows poor burning and starting characteristics and loud humming ballasts.
3. *SWITCHES:* Defective operation broken or missing parts, arcing noises.
4. *CONVENIENCE OUTLETS:* Dirty, inadequate defective contacts, difficult plugging, overheating, evidence of overloading or multiple sockets servicing lamps or appliances, lack of grounding terminal.
5. CORDS, CORD EXTENSIONS, AND PORTABLE APPLIANCE CORDS:
   a. Inadequate, unsafe, unreliable, incorrect types being used.
   b. Lengths too long, poor insulation, twisted, exposed to damage underfoot, spliced, lying on floor or across heated surfaces or lamps, lamp types used for portable extensions that are subject to moisture, oil, and grease.
   c. *Plugs:* cracking, breaks, loose connections, wires improperly attached and in danger of pulling away from plug when removing from outlet, missing protective cover on male ends, no grounding terminal or ground wire with clamp.
6. *LIGHTING VOLTAGE:*
   a. Spot measurement at fixture indicates measured voltage of nominal lamp volt-

E	U	R	D

ages, and lighting outlets in excess of 6% of nominal lamp voltage.

b. Unauthorized connections of hot plates, coffee pots, heating devices, and other electrical equipment on lighting circuits.

c. Interference with branch circuits for power and lighting from motor starting or stopping, such as light flicker, or excessive voltage dips causing fluorescent and mercury lamps to drop out.

7. *ILLUMINATION LEVELS:*

a. Ambient conditions such as dirty walls and ceilings

b. Spot measurement of light levels using accurate foot candle meter indicates depreciation of 20 to 25% of level obtainable from clean fixtures and new lamps.

c. Failure to keep continuous record of illumination levels at established check points.

Total Items: 7 Raw Total %

Finding: (comment on each checkpoint and summarize to justify scoring):

____

____

____

____

____

____

____

E	U	R	D

Section 16. Switchgear

SCOPE: Electrical switchgear, associated apparatus, and equipment connected to distribution circuits in buildings, 600 volts and under, not located in vaults or fire-resistant rooms. Included are primarily switchgear known as "metal clad," "drawout," "cubicle," or "truck" type.

1. *HOUSEKEEPING:*
   a. Poor appearance, dampness, dirty, inaccessibility of surrounding areas.
   b. Detrimental conditions such as ambient temperatures in excess of 100°F; humidity causing sweating of metal enclosures; rodents, insects; stored combustibles; trash, dirt, dust accumulations; poor location, inaccessibility; poor ventilation; gas, steam, or water leakage.
2. *EXTERIOR HOUSING AND ENCLOSURE GROUND:*
   a. Rust, corrosion, need for painting, signs of abuse, unauthorized or improper signs, scribbling, calendars, storage of materials or dust accumulations on top of enclosures, missing parts, schedules, or other items.
   b. Poor condition and inadequacy of enclosure ground.
3. *INTERIOR OF COMPARTMENTS, CUBICLES, AND DRAWERS:* Dirty, condensation, symptoms of overheating, burns from grounds and short circuits, defective insula-

E	U	R	D

tion, insulation, defective operation of locks, doors, and drawers.

4. *AIR AND OIL CIRCUIT SPEAKERS, OIL-LESS-TYPE AIR BLAST BREAKERS (DE-ENERGIZED (50 Amperes and above)*:
   a. Incorrect wipe of main and arcing contacts on opening and closing: measure actual contact impression length with indicating paper. Carefully dress silver contacts to original contours to provide minimum contact length of ¾ of length of contact plates.
   b. Overheating, lack of continuity, and looseness of connections on all mechanisms; incorrectly placed pins and cotter pins.
   c. Improper functioning of rods and moving parts and binding when breaker is operated. Ensure that bearing points are lubricated and clean, particularly near contacts and arc quenchers.
   d. No freedom or action in tripping devices and latches; insufficient travel of armature overcurrent trip devices to ensure release of breaker latch.
   e. Check dirty oil in oil-film timers.
   f. Improper functioning and unsatisfactory condition of control switch and closing relay.
   g. *Oil Tanks:* Check for leaks, cracks, corrosion, defective gaskets, improper oil levels, incorrect gauge indications, oil for presence of dirt and sludge. Test oil.
   h. *Bushings and Insulators:* Check accumulation of dirt, cracks, chips, lack of rigidity of supports, inadequacy of connections.
   i. *Arc Chutes:* Check for moisture and other contaminants, arc mufflers for loose scale, runners and springs for wear or other damage.

E	U	R	D

j. Determine if indicating lamps are operative.

5. *PNEUMATIC SWITCHES:* Poor general condition and unsatisfactory operation of all components serving breakers, including air pressure valves, pistons, and associated equipment; poor speed of air pressure recovery, air leaks, moisture and dirt in air lines.
6. *ENCLOSED SWITCHES:*
   a. Unsatisfactory operation of handle and mechanism: blade latch functions improperly; blades are insecure when closed.
   b. Blade, clips or tongue for lack of cleanliness, signs of oxidation, arcing, overheating, and poor contact. (A 0.002-inch feeler gauge inserted between blade and clip or blade and tongue will not go into a good bright contact.)
7. *SWITCHGEAR RELAYS:* Visible evidence of accumulations of dust and dirt, broken glass, loose or missing nuts, loose mounting, noisy operation, signs of arcs, loose connections.
8. *DRY-TYPE TRANSFORMERS:*
   a. Accumulation of dust on core, coils, and leads; evidence of moisture, oil, or other contaminants; evidence of overloads and burned insulation; poor contacts and connections; signs of corrosion; poorly painted surfaces, nameplates, guards and warning signs; inadequate ventilation, detrimental ambient temperatures, and defective ground connections
   b. Unsatisfactory operation of air-temperature alarms, auxiliary cooling equipment, and tap connections.
   c. When operating log is kept, check operating temperature loads and voltage during critical operating periods from log data.

E	U	R	D

d. Detrimental ambient conditions, defective lighting protection, moisture, overloads, physical injury or abuse, and short circuits causing insulation deficiencies.

9. *INSTRUMENT TRANSFORMERS: Lack of cleanliness, inadequate insulation connections, inadequate insulation, improper ventilation, and unsatisfactory grounding.*
10. *SMALL WIRING:* Physical damage, defective insulation, poorly soldered connections, damage or other connections.
11. *BUSES AND BUSWORK:* Lack of cleanliness, defective insulation overheating, signs of flashovers.
12. *RHEOSTATS AND ASSOCIATED MECHANISMS:* Inadequate operation, lack of cleanliness, overheating, dirty contacts, overtravel.
13. *MECHANICAL DEVICES: DEENERGIZED:* Unsatisfactory operation and inadequacy of mechanical devices for elevating and lowering or drawing out switchgear; improper operation of interlocks, gang switches, and contact safety shutters.
14. *SWITCHGEAR FOUNDATIONS:* Settling or movement of floors, pedestals, and other foundations, resulting in possible misalignment of operating parts in switchgear.
15. *TESTS:*
    a. Measure operating voltage on electrically operated breakers (see Table 2.1 below).
    b. Measure operating voltages in switchgear control circuits.
    c. Make liquid dielectric strength tests. Recommend that liquid tests under 18,000 volts be changed.
    d. Measure insulation resistance and evaluate: minimum value or low limit recommended is 2 megohms total (after one

TABLE 2.1 Operating Voltage Range

Nominal Control Voltage	Closing coils of solenoid mechanisms, motors of motor mechanisms, operating coils of pneumatic mechanisms	Trip Coils—all mechanisms
48 DC	36–53	28–60
125 DC	90–130	70–140
250 DC	180–260	140–280
115 AC	95–125	95–125
230 AC	190–250	190–250

E	U	R	D

minute at approximately 25°C). For safe operation or before making high-potential tests: make high-potential tests at 65% of initial installed value or check manufacturer's recommendation.

Total Items: 15 Raw Total %

Findings (comment on each checkpoint and summarize to justify scoring):

____

____

____

____

____

____

E	U	R	D

Division

Operational Facilities Assessment

E	U	R	D

I. OBJECTIVE: The basic objective is to maintain operational facilities in an economical manner that will protect the facility's investment, reduce hazards to life and property, and permit continued service consistent with the facility's mission.

II. DEFINITIONS

A. Operational facilities are classified as follows:

1. Towers, masts, antennas
2. Chemical/fuel facilities
3. Waterfront: Brows, gangways, camels, separators
4. Electrical distribution: Disconnecting switches, distribution transformers, electrical grounds and grounding systems, electrical instruments, electrical potheads, electrical relays, lighting arrestors, power transformers, safety fences, vaults and manholes, poles and accessories, and substations

E	U	R	D

5. Electrical utilization: cathodic protection systems, electric motors and generators, fuses and small circuit breakers, pier circuits and receptacles, rectifiers, telephone substations, and street/flood/security, lighting

B. Sections. Sections for which inspection checklists are established are depicted above.

III. MAINTENANCE STANDARDS: The degree of maintenance, repair, and rehabilitation of operational facilities shall be governed by known foreseeable usage. All items in this category shall be maintained to the extent necessary to ensure reliability of service, full safety of operators, efficient operation of the equipment/item, and the prevention of unwanted deterioration of equipment.

Section 1 Towers, Masts, and Antennas

SCOPE: Antenna-supporting towers and masts, guyed radiators and guys, antenna-supporting strongbacks, strongback insulation to towers, elevating mechanisms, and obstruction and navigation lighting. These checkpoints do not cover the inspection of antennae or beacons.

1. FOUNDATIONS: Check for cracked, broken, or spalled concrete; exposed reinforcing bar; movement or settlement; heaving from frost action.
2. ANCHOR BOLTS AND STRAPS: Check for rust, corrosion, and loose, missing, or damaged parts.
3. STRUCTURAL STEEL TOWERS, LADDERS, AND SAFETY CAGES: Check for rust, corrosion, loose, missing, twisted, bowed, bent, or broken members.

	E	U	R	D
4. SPLICES, BOLTS, AND RIVETS: Check for rust, corrosion, loose, missing, other damage, broken welds.				
5. TIMBER TOWERS AND LADDERS: Check for loose, missing, twisted, bowed, cracked, split, rotted timber; termite or other insect infestation of members.				
6. GUYS AND ANCHORAGE: Check for cracked, split, rotted; termite or other insect infestation, or looseness of wooden parts; metal parts for rust, corrosion, loose, missing. or other damage, frayed or broken strands and looseness of guys; inadequate deadman anchorage.				
7. PLUMBNESS: Check for any deviations.				
8. PAINTED SURFACE: Check rust, corrosion, cracking. scaling, peeling, wrinkling, alligatoring, scaling, fading, complete loss of paint.				
9. OBSTRUCTION AND NAVIGATION LIGHTS: relamping.				
10. LIGHTS: Check for improper operation and lack of cleanliness of lights, shields, hoods, and receptacle fittings.				
11. STRIKING, BINDING, ARCING, OR BURNING OF RELAY CONTACTS: Check for loose connection, missing parts of relays.				
12. LIGHTNING RODS AND AERIAL TERMINALS: Check for damage from burning.				
13. CONDUITS, TERMINALS, AND DOWN-LEADING CABLES: Check for corrosion, structures, loose or missing attachments to structures, other damage.				
14. Check for poor or unsatisfactory mechanical bonding of joints or aerial terminals, down-leading cables, and ground connections.				
15. ELECTRICAL CONTINUITY: Check for lack of electrical continuity from aerial terminals through ground connections.				

E	U	R	D

16. INSULATORS: Check for dirt, dust, grease, or other deposits on insulators, or cracks, breaks, chips, or check of the porcelain glaze.
17. PULLEYS, WINCHES, CABLES, ROPES, OR OTHER ELEVATING MECHANISMS: Check for inadequacy or improper operating condition.

Total Items: 17 Raw Total %

Findings (comment on each checkpoint and summarize to justify scoring):

____

____

____

____

____

____

____

____

Section 2 Chemical/Fuel Facilities (Receiving and Issue)

SCOPE: This section covers platforms and islands, small structures, fuel hoses, hose connections and adapters, hose racks and reels, grounding connections, portable ladders and steps, portable gangplanks, signs and markings, general cleanliness, and painting. Pumps, valves, piping, tanks, security fences, derricks, railway

E	U	R	D

trackage, concrete and surfaced roadways, piers and wharves, electric motors and their control equipment, and ground connections for electrical equipment are covered in other sections.

1. WOODEN PLATFORMS AND ISLANDS: Check for loose, missing, worn, rotted, other damage to individual planks.
2. WOOD FRAMING, SUPPORTS, STAIRS, AND GUARDRAILS: Check for loose, missing, worn, rotten, broken parts.
3. METAL FRAMING, SUPPORTS, STAIRS, AND GUARDRAILS: Check for rust, corrosion, loose, missing, twisted, bowed, bent, broken parts.
4. METAL PLATFORMS: Check for worn, bent, broken, or otherwise defective gratings and plates.
5. CONNECTIONS: Check for rust, corrosion, loose, missing, broken, other damage.
6. CONCRETE ISLANDS: Check cracks, breaks, settlement that may lead to failure of pipe, valves, and
7. SMALL STRUCTURE CONCRETE FOUNDATIONS: Check for breaks, settlement that may lead to failure.
8. METAL AREAS: Check for rust, corrosion, wear, other damage.
9. WOOD FRAMING: Check for wear, rot, insect infestation, other damage.
10. ROOFS AND/OR WALLS: Check for leakage, wear, rot, insect infestation, other damage.
11. DOORS AND WINDOWS: Check for sagging, binding.
12. HARDWARE: Check for defective hinges, locks, broken glass.
13. HOSE RACKS AND REELS: Check for corrosion of metal, rotting and other damage to wood, mechanical damage, or other defects that may result in injury to stored hoses.

E	U	R	D

14. PORTABLE LADDERS, STEPS AND GANGPLANKS: Check for wear, mechanical damage, breakage, loose, missing, or damaged bolts or connections, other defects that may be a safety hazard.
15. SIGNS AND MARKINGS: Check for inaccurate and illegible.
16. DEBRIS AND SPILLAGE: That may cause safety or tire hazard.
17. PERMANENT GROUNDING CONNECTIONS FOR STEEL STRUCTURES, PIPING, AND RAILWAY TRACKAGE AT FUEL PIERS AND AT RECEIVING AND ISSUE STANDS: Check for mechanical and corrosive damage.
18. PORTABLE GROUNDING CONNECTIONS: Check for mechanical and corrosive damage.
19. TEST GROUNDING CONNECTIONS: Check for permanent and portable grounding connections with megger or ohmmeter to ensure electrical continuity and zero grounding.
20. PAINTED SURFACES: Check for rust, corrosion, cracking, scaling, peeling, wrinkling, alligatoring, chalking, fading, or complete loss of paint.

Total Items: 20 Raw Total %

Findings: (comment on each checkpoint and summarize to justify scoring):

____

____

____

____

E	U	R	D

Section 3. Chemical/Fuel Facilities (Storage)

SCOPE: Surface and subsurface tanks, tank enclosures, and tank fittings and appurtenances.

1. FOUNDATIONS: Check for settling, movement, upheaving, inadequate soil coverage.
2. EXTERIOR CONCRETE SURFACES: Check for spalling. cracking, exposed reinforcing bar, leakage.
3. EXTERIOR STEEL SURFACES: Check for rust, corrosion, distortion or other structural failure, leakage, deteriorated paint.
4. ROOF SURFACES: Check for defects in waterproofing, heat reflecting coatings, coverings.
5. FLOATING AND EXPANSION-TYPE ROOFS, SEALS, SUPPORTS, AND SUPPORT GUIDES: Check for rust, corrosion, improper sealing, deteriorated paint, structural or mechanical damage caused by freezing weather conditions.
6. STRUCTURAL SUPPORTS AND CONNECTIONS: Check for rust, corrosion, rot, broken, cracked, distorted, loose, missing, deteriorated paint.
7. TANK LININGS: Check for loss of elasticity, granulation, discoloration, cracks, peeling, sloughing off.

Item	E	U	R	D
8. TANK INTERIOR: Check for rust, corrosion, scale, deteriorated protective coatings.				
9. FRAMES AND COVERS ON MANHOLES AND HATCHES: Check for rust, corrosion, cracks, breaks, missing or damaged bolts, worn or defective hinges and gaskets.				
10. VENTS: Check for rust, corrosion, dirty screens.				
11. PRESSURE AND VACUUM RELIEF VALVES: Check for defective operation, leakage, loss of fluid.				
12. MANOMETERS AND THERMOMETERS: Check accuracy, mechanical damage, loss of fluid.				
13. FLOAT GAUGES: Check for wear, binding, apparent inaccuracy.				
14. CABLES, SHEAVES, AND WINCH OF SWING LINES: Check for wear; mechanical damage; stuffing boxes and liquid seals; deterioration; improper operation.				
15. STAIRS, LADDERS, PLATFORMS, AND WALKWAYS: Check for rust, corrosion, rot; broken, cracked, loose, missing, members or connections; deteriorated paint.				
16. ROOF DRAINS AND SCREENS: Check for missing, rust, clogging.				
17. GROUND CONNECTIONS: Check for loose, missing, mechanical damage; corrosion interfering with electrical continuity.				
18. INTERIOR HEATING, INLET AND OUTLET PIPES, NOZZLES SUPPORTS, SUMPS, AND SUMP DRAINS: Check for rust, corrosion, wear, loose or missing parts, obstructions. other defects.				
19. DIKES: Check for cracks, breaks, spalling, rust, corrosion, settlement, heaving, soil erosion, water seepage; inadequate sod cover on outer face where earth filled; inadequate treatment of inner face to prevent vegetation				

E	U	R	D

growth; access steps for settlement, breaks, other damage.

20. DRAINAGE DITCHES, SUMPS, AND EARTH SURFACES BETWEEN DITCH AND FOUNDATION: Check for proper slope to divert surface water away from foundation and berm; trash and debris; erosion.
21. LEAKAGE: Review records to determine fuel losses; water in oil samples. Locate leaks by filling completely with water and applying hydrostatic pressure of 4 feet of water for not less than 4 hours.

Total Items: 21 Total Raw %

Findings (comment on each checkpoint and summarize to justify scoring):

____

____

____

____

____

____

____

Section 4. Brows and Gangways

SCOPE: Brows and gangways constructed of steel, wood, and aluminum.

1. Abnormal deflection.

E	U	R	D

2. STEEL AND ALUMINUM STRUCTURAL MEMBERS: Check for rust, corrosion, loose, missing, bent, buckled, or broken members.
3. STEEL AND ALUMINUM FLOORING MEMBERS: Check for mechanical damage, excessive wear, slippery walking surfaces.
4. HANDRAILS (OR CABLES) AND STANCHIONS: Check for damaged or loose members and fittings; wire rope for looseness, frayed or broken strands.
5. ROLLERS OR WHEELS: Check for ease of movement, lack of lubrication.
6. CONNECTIONS: Check for rust, corrosion, loose, missing, broken, other damage.
7. WOODEN STRUCTURAL MEMBERS: Check for loose, missing, twisted. bowed, cracked, split, broken, or rotted members, particularly at joints.
8. WOODEN FLOORING MEMBERS: Check for mechanical damage, excessive wear, slippery walking surfaces.
9. HANDRAILS: Check for loose, missing, or broken wooden handrails.
10. PAINTED SURFACES: Check for rust, corrosion, cracking, scaling, peeling, wrinkling, alligatoring, chalking, fading, complete loss of paint.

Total Items: 10 Raw Total %

Findings (comment on each checkpoint and summarize to justify scoring):

____

____

____

E	U	R	D

Section 5. Camels and Separators

SCOPE: Camels and separators constructed of timber and steel.
SECURING CHAINS (OR LINES) FITTINGS AND HARDWARE: Check for broken, loose, or missing parts; excessive wear, rust, corrosion.

1. TIMBER:
   a. Missing or broken members.
   b. Excessive damage, severe wear, external decay.
   c. Evidence of internal decay, attack by marine borers,deterioration.
   d. Individual timbers: lack of buoyancy, deterioration in cross-section area of 50% or more.
2. STEEL PONTOONS:
   a. Missing units.
   b. Mechanical damage, excessive wear.
   c. Rust, corrosion, lack of paint.

E	U	R	D

d. Individual units: lack of buoyancy, reduction in shellthickness of 40% or more.

Findings: (comment on each checkpoint and summarize to justify scoring):

____

____

____

Section 6. Dolphins

SCOPE: Multiple timber pile dolphins.

1. Missing, loose, or broken piles.
2. Splitting, deep abrasion, excessive wear.
3. Damage from marine borer activity or decay.
4. Report for replacement piles when cross-section area has deteriorated 30% or more.
5. Vertical misalignment of dolphin assembly.
6. Loose, missing, broken, or excessively worn or corroded cables.
7. Loose, missing, broken, or corroded cable connections.
8. Loose, missing, broken, or deteriorated wedge blocks.
9. Loose, missing. broken, or corroded wedge block holding bolts.
10. Loose, missing, broken, or corroded chafing strip and bands.
11. Loose, missing, broken, or corroded supporting check bolt hangers.
12. Need for rust removal and application of protective coatings.

Total Items: 12 Raw Total %

E	U	R	D

Findings (comment on each checkpoint and summarize to justify scoring):

____

____

____

Section 7. Piers, Wharves, Quaywalls, and Bulkheads

SCOPE: Piers, wharves, quaywalks, and bulkheads constructed of timber, steel, or concrete.

1. HORIZONTAL MISALIGNMENT (USE SURVEYOR'S TRANSIT): Check for outward movement indicated by increase in expansion joint width or cracks in adjacent ground or pavement areas.
2. VERTICAL SETTLEMENT (USE SURVEYOR'S LEVEL): Check for notable settlement.
3. CURBINGS, HANDRAILS, AND CATWALKS: Check for loose, missing, broken individual sections; uneven surfaces, obstructions, other hazardous conditions.
4. BOLLARDS, BITS, CLEATS, AND CAPSTANS: Check for breakage; excessive wear; rough or sharp surfaces and edges detrimental to handling lines; missing, loose, or defective bolts.
5. DECK DRAINS AND SCUPPERS: Check for loose, missing, or broken screws; water ponding, indicating clogged drains or need of deckfill or other repair.
6. MANHOLE COVERS AND GRATINGS: Check rust, corrosion, mechanical damage, bent or worn hinge pins.

	E	U	R	D
7. ASPHALT DECK COVERING: Check for cracks, weathering, holes shoving, rutting, or porous surfaces.				
8. LADDERS: Check for rust, corrosion, loose, missing, twisted, bowed, bent, or broken steel members; loose, missing, cracked, split, rotted, or broken wooden members; defective supports and anchorage.				
9. DECK PLANKING: Check for worn, loose, missing, cracked, broken, splintered, deep abrasions, decay; termite and other pest infestation, particularly on underside surfaces.				
10. WOOD STRINGERS AND PILE CAPS: Check for loose, missing, cracked, broken, decay; termite and other pest infestation; crushing of wood fiber, particularly at bearing points; ineffective bearing area not in contact with bearing piles.				
11. WOOD BEARING PILES: Check for missing, broken, loss of section, porosity of surface; pest infection, and decay, particularly at top and areas within tide range.				
12. WOOD FENDER PILES: Check for loose, missing, cracked, broken; excessive wear; mechanical damage; termite and other pest infestation; wrappings for broken loose, or missing cables and fastenings; loose or missing wedges.				
13. WOOD, BRACINGS, WALES, AND CHOCKS: Check for loose, missing, broken, split, warped, decay, termite and other pest infestation, rotted bolt holes.				
14. Report for replacement individual wooden members when deterioration has reduced cross-section area by 30% or more, except 50% or more for bracings, wales, and chocks.				
15. FIREWALLS: Check for structural damage, general deterioration, lack of airtightness.				

E	U	R	D

16. STEEL DECK PLATES, GRATINGS, AND THEIR SUPPORTS: Check for rust, corrosion, loose, missing, broken, bent.
17. STEEL BEAMS, GIRDERS, AND PILING: Check for rust, corrosion, structural damage, misalignment, defective connections.
18. STEEL TIE AND ROD BRACINGS, LONG BOLTS, AND WALES: Check for rust, broken, bent; loose nuts and anchor plates; lack of tautness; failure or missing wrappings on tie rods; defective connections.
19. Report for replacement or repair individual steel members when deterioration has reduced flange or web thickness by:
    a. 50% or more for rods and long bolts.
    b. 40% or more for bracings and wales.
    c. 30% or more for all other members.
20. GROUNDING CONNECTIONS: Check for loose, missing, rust, corrosion, mechanical damage, electric discontinuity.
21. CONNECTIONS: Check for rust, corrosion, loose, missing, broken, other damage.
22. Report steel for cleaning and painting where surface area is corroded:
    a. 50% or more on deck plates, piles, bracing, and wales.
    b. 40% or more on individual beams and girders.
23. Report for cleaning and painting all steel beams and girders in those spans where total surface area of beams and girders in one or more adjacent spans is corroded 50% or more.
24. CONCRETE SLABS: Check for general deterioration, breakage, cracking, spalling, particularly at bottom surfaces; reinforcing steel for exposure, rust, corrosion; expansion joints for loose or missing filler. Report reinforcing for repair when broken or cross-section area reduced 40% or more.

E	U	R	D

25. CONCRETE BEAMS AND GIRDERS: Check for breakage, cracking, spalling, particularly at edges; reinforcing steel for exposure, rust, corrosion.
26. CONCRETE PILES: Check for breakage, cracking, spalling, particularly within areas of tide range; reinforcing steel for exposure, rust, corrosion.
27. BULKHEADS: Check for loss of section; excessive weathering; loss or settlement of backfill; erosion of bottom, check depth of water adjacent to bulkhead for comparison with original depths.
28. BULKHEAD TIE-RODS: Check for loose or missing connections; corrosion.
29. PROTECTIVE COATED AREAS: Check for rust, corrosion, deteriorated coating, complete loss of coating.
30. Accumulations of combustible materials, debris, or drift materials.

Total Items: 30 Raw Total %

Findings (comment on each checkpoint and summarize to justify scoring):

____

____

____

____

____

____

____

E	U	R	D

Section 8. Disconnecting Switches

SCOPE: Manually group-operated and hook-stick-operated disconnecting switches used on transmission lines and distribution systems, including grounding switches.

1. OPERATING GEAR:
   a. Group-operated switches: check for rust, corrosion, loose brackets and holding bolts, nonrigid bearings and supports.
   b. Grounding cables, clamps, and straps: check for weak supports, broken or frayed portions of conductors, loose connections.
   c. Insulating sections of operating rod: check for indications of cracks or signs of flash-overs.
   d. Movable connections: check for inadequate lubrication, rust, corrosion, other conditions resulting in a malfunction.
   e. Switch: check gears for stiffness or adjustment needed, switch several times to determine. (DO NOT operate without prior clearance.)
   f. Locking and interlocking devices and mechanisms: check for functional inadequacy to prevent unauthorized operation.
2. MOUNTINGS AND BASES: Check for rust, corrosion; twisted, bent or warped; loose or missing ground wire.
3. INSULATORS:
   a. Check for cracks, breaks, chips, chiping of porcelain glaze; more than thin or transparent film of dirt, dust, grease, or other deposits on porcelain.

E	U	R	D

b. Check for damage indicated by streaks or carbon deposits from flashovers.
c. Check for loose, broken, or deteriorated cement holding insulator to other parts.

4. BLADES AND CONTACTS:
   a. Excessive discoloration from overheating: Check for roughness and pitting from arcing.
   b. Check for misalignment of blades with contacts.
   c. Arcing horn contacts: burns, pits, failure to contact each other throughout their length when switch is opened and closed.
   d. Inadequate tension of bolts and springs.
   e. Inadequate blade stop.
   f. Lack of hinge lubrication: Check sufficiency of non-oxide grease for blades and contacts.
5. CONNECTIONS:
   a. Cable or other electrical connections: Check for discoloration, indicating excessive heating at connection points.
   b. Check for corrosion, particularly that resulting from atmospheric conditions.
   c. Electrical clearances of cable or other conductor: Check for adequacy of other phases or to ground for applicable circuit voltage. (Switch both open and closed.)
   d. Flexible connections: Check for frayed, broken, or brittle. (Excessive discoloration indicates overheating.)
   e. Cable from grounding switch to grounding system: Check for frayed, broken strands, loose connections.

Total Items: 5 Raw Total %

Findings (comment on each checkpoint and summarize to justify scoring):

E	U	R	D

Section 9. Electrical Grounds and Grounding Systems

SCOPE: Electrical grounds and grounding systems for all electrical equipment, apparatus, machinery, metallic conduit, and all items that are a part of the outdoor electrical power distribution system. Also included are grounds of structural supports, frames, towers, safety fencing, hardware, equipment enclosures, system neutrals, and buried ground cable networks. Good engineering practice, and controlling rules and regulations such as National Electric Code, require grounding for operational and personnel safety.

1. Visual connections: Loose, missing, broken connections; signs of burning or overheating, corrosion, rust, frayed cable strands, more than 1 strand broken in 7-strand cable, more than 3 strands broken in 19-strand cable.
2. Underground connections: Unsatisfactory condition or defects uncovered when 4 or 5 connections are exposed to view by digging.

E	U	R	D

Maximum permissive resistance for grounds and grounding systems between equipment or structure being grounded and solid ground (earth) see Tables 2.2 and 2.3.

Total Items: 2 Raw Total %

Findings (comment on each check point and summarize to justify scoring):

____

____

____

____

____

____

Section 10. Electrical Instruments

SCOPE: Electrical AC and DC indicating recording, and portable instruments and associated equipment used for measurement of electrical power quantities.

1. GENERAL:
   a. Improper shieldings, mounting, or enclosures when located near strong magnetic fields, subject to vibrations, extremes in temperature, moisture, metallic and other dust, and acid or corrosive vapors.
   b. Inadequately, improperly, not neatly stored in cases or cabinets that are free from dust, corrosive fumes, excessive heat, moisture, vibration, strong magnetic fields.

TABLE 2.2 Grounding Systems

From	To	Allowable Resistance
a. Point of connection structure, equipment, enclosure, or neutral conductor	Top of ground rod	See Table 2.3
b. Ground rod, mat, or network	Ground (earth)	See Table 2.3
c. Gates	Gateposts	½ OHM
d. Operating rods and handles of group-operated switches	Supporting structure	½ OHM
e. Metallic-cable sheating	Ground rod, cable, or metal structure	½ OHM
f. Equipment served by rigid conduit	Nearest ground	10 OHMS

TABLE 2.3

Allowable Resistance	
Grounding System	Maximum Permissible Resistance (ohms)
1. For generating stations:	1
2. For main substation, distribution substations, and switching stations on primary distribution systems	3
3. For secondary distribution system (neutral) grounding, non-current-carrying parts of distribution system itself, enclosures of electrical equipment not normally within reach of other than authorized and qualified electrical operating and maintenance personnel	10
4. For individual transformer and lightning-arrester grounds on distribution system	10
a. When total resistance in check point 3 or 4 exceeds allowable, measure resistances of individual portions of the circuits to determine the points of excessive resistance and report.	
b. Substandard resistance values resulting from poor contact between metallic portions of grounding system and earth.	
c. Structural steel, piping, or conduit run exceeding 25 feet used as a current-carrying part of grounding circuit for protection of equipment.	
d. Absence of ground-cable connections.	

E	U	R	D

c. Not clean, improperly marked and identified, incorrect type and range for application, no manufacturers' instructions for servicing.

d. Poor physical condition of instrument cases, portable cases, handles, nameplates, leads, calibrated leads, shunts, multipliers.

e. Loose electrical connections; dirty or corroded contact surfaces; inadequate, poorly arranged, improperly insulated wire, cable, and leads.

f. Broken glass, pointer friction, warped or dirty or scale, bent pointers, and missing parts.

g. Moving elements not locked when instruments provided with locking devices are not in use.

h. Inkwell not clean and dry when portable type recording instruments are in storage.

2. CALIBRATION: Not serviced, calibrated, or tested at appropriate intervals to accepted standard of accuracy for particular instrument; records of tests not available.
3. WATT-HOUR METERS:
   a. Nonoperative voltage (power-on) indicating lamps.
   b. Outdoor service meters: poor physical condition, loose weather seals, moisture or dirt in enclosure, corrosion, loose connections, missing parts.
4. DC AMMETERS: Improperly connected in grounded leg of grounded DC circuits.
5. INSTRUMENT TRANSFORMERS: Poor physical condition, dirty, inadequate connections, visual evidence of overloading or overheating.

Total Items: 5 Raw Total %

E	U	R	D

Findings (comment on each checkpoint and summarize to justify scoring):

Section 11. Electrical Potheads

SCOPE: Electrical potheads used in power distribution systems, including potheads used as terminals of underground cables as well as those incorporated as terminals of equipment and a part thereof.

1. PORCELAIN:
   a. Check for cracks, breaks, chips, checking of porcelain glaze.
   b. Check for streaks of carbon deposits, indicating flashovers and possible damage.
   c. Check for dirt, dust, grease, other deposits.
   d. Check for cracks, breaks, or deterioration of cement sealing compound, and leakage or signs of moisture.

E	U	R	D

2. CABLE CLAMPS: Check for corrosion, loose bolts, solder, poor mechanical connections. (Corrosion of lead cables and connections at pothead indicated by presence of a white, brownish, or reddish product.)
3. TERMINAL STUDS AND BOLTING PADS: Corrosion, loose connections, and poor contacts evidenced by discolorations from heating.
4. MOUNTINGS: Check for corrosion and other weaknesses.

Total Items: 4 Raw Total %

Findings (comment on each checkpoint and summarize to justify scoring):

____

____

____

____

____

____

____

____

Section 12. Electrical Relays

SCOPE: Protective relays located in electrical power circuits only. It does not include relays used to protect or control utilization equipment. Relay adjustments, settings, and electrical tests

E	U	R	D

are not covered. Assessment is limited to those aspects that relate to care rather than to the operating characteristics of relays.

1. GENERAL:
   a. Check for dirt, evidence of moisture, high temperature, and other adverse conditions.
   b. Check for visible corrosion, deterioration, or pitting of contacts, pivots, and coils.
   c. Check for broken or loose parts and connections.
2. TEMPEPATURE AND PRESSURE RELAYS:
   a. Check settings for improper temperature and pressure limits.
   b. Check for evidence of damaging temperature or pressure conditions.

Total Items: 2 Raw Total %

Findings (comment on each checkpoint and summarize to justify scoring):

____

____

____

____

____

____

____

____

Section 13. Lightning Arresters

E	U	R	D

SCOPE: All types of lightning arresters for protection of electric power distribution lines and equipment. The types include the treated-ceramic-gap type, such as Thyrite or Autovalve, the oxide-film type, the obsolete pellet type, used for distribution systems and lower-voltage transformer protection (up to 34.5 kv), the capacitor type, used mostly for protection of rotating equipment, and the expulsion-gap type, used to reduce outages from flashover caused by lightning.

1. FOUNDATIONS AND SUPPORTS: Signs of weakness, cracked or broken concrete, burns, loose hold-down bolts, rust, mechanical damage.
2. GROUNDING CABLES FOR POLE-MOUNTED LIGHTNING ARRESTERS: (where Accessible to public): Check for cracks, breakage, splintering, defective paint, evidence of tampering, other weakness in protective moldings.
3. TREATED CERAMIC-GAP TYPE:
   a. Porcelain insulators: check for signs of flashovers and serious flashover marks; scarring, chipping, or cracking of porcelain; dirt, grease, or other film on porcelain.
   b. Metal bases, caps, and intermediate section couplings: check for indications of loose bolts, rust, corrosion, or loose cement.
   c. Connections to line, equipment, or ground lead: check for looseness, corrosion, breakage, or misalignment that may put undue mechanical strain on porcelain.
   d. Ground cable connection to ground mat: check for loose or corroded connectors where visible.
4. OXIDE-FILM TYPE:

Item	E	U	R	D
a. Check for accumulation of dirt, particularly on edges of cells, deterioration of paint, rust, corrosion.				
b. Check for loose connection and mounting bolts, badly corroded connection posts.				
5. PELLET-TYPE (obsolete, not acceptable for replacement):				
a. Check for indications of burns and scars on porcelain bodies from flashovers, cracked or broken bodies and caps.				
b. Mounting Clamps: Check for rust, corrosion, loose bolts at arrester and at supporting point of bracket.				
c. Check for poor physical condition of ground cable from arrester to point of connection to ground rod or grounding system, where visible; loose or corroded connectors.				
6. CAPACITOR-TYPE:				
a. Check for signs of flashover on porcelain insulators and metal enclosures resulting in cracking, breaking, or burning.				
b. Connection points: Check for looseness, corrosion, frayed ground cables, evidence of mechanical strain.				
c. Enclosures: Check for excessive rust and corrosion.				
d. Porcelain: Check for dirt accumulations in appreciable amounts.				
7. EXPULSION-GAP TYPE:				
a. Check for looseness of mounting, flashovers, damage to tubing, corrosion, loose ground connections, signs of burning and apparent damage from visual check of gap opening between arcing horn and line being protected.				
b. Check for poor physical condition of ground cable from arrester to point of connection to ground rod or grounding system; loose or corroded connectors.				

E	U	R	D

c. Check for signs of burning on external air gaps.
d. Check opening of air gap; examine tube carefully for damage from flashovers and burnouts; check for corrosion of metal mounting parts.

Total Items: 7 Raw Total %

Findings (comment on each checkpoint and summarize to justify scoring):

____

____

____

____

____

____

____

____

____

____

____

____

E	U	R	D

Section 14. Power Transformers, Deenergized

SCOPE: Deenergized electric power transformers used for voltage transformation on transmission lines and high voltage distribution systems. Before inspection, make arrangements to have electricians and other required labor available.

1. BUSHINGS AND INSULATORS:
   a. Insulators and porcelain parts: Check for indications of cracks, checks, chips, breaks; where flashover streaks are visible, reexamine for injury to glaze or for presence of cracks.
   b. Check for chipped glazing exceeding ½ inch in depth or an area exceeding 1 square inch on any insulator or insulator unit, report for investigation by a qualified electrical engineer.
   c. Check for severe cracks, chipped cement, or indications of leakage around bases of joints of metal where it connects to porcelain parts at terminal and transformer ends.
   d. Terminal ends: Check for mechanical deficiencies, looseness, corrosion, damage to cable clamps.
   e. Check for improper oil level in oil-filled bushings.
   f. Check for heating as evidenced by discoloration and corrosion indicated by blue, green, white, or brown corrosion reidue, on metallic portions of all main and ground terminals, including terminal board and grounding connections inside transformer case.
   g. Pipe, bar copper. and connections: Check for indications of overheating or flashover fusing.

Item	E	U	R	D
h. Cable connections: Check for broken, burned, corroded, missing strands. (Fused portions of connectors, cables, pipe, or bus copper should be filed smooth and all projections removed; clean metalwork, disconnecting if required and cover with thin coating of nonoxide grease; if connections are disassembled, rough spots on contact surfaces should be filed smooth and all projections removed; see that all bolted and crimped connections are tight by setting up nuts or recrimping when looseness is evident or suspected; clean and tighten all corroded or loose connections; repair or replace cables with frayed and broken strands: repair frayed or broken cable insulation.)				
2. ENCLOSURE AND CASES: If case is opened for any reason, examine immediately for signs of moisture inside cover, and where present, for plugged breathers, inactive desiccant, enclosure leakage.				
3. COILS AND CORES: When cover is open, examine interior for deficiencies, dirt, sludge. If feasible, probe down sides with glass rod, and if dirt and sludge exceeds approximately ½ inch, arrange to have filter insulating oil changed and have coils and cores cleaned.				
4. BUSHING-TYPE INSTRUMENT TRANSFORMERS:				
a. Indications of deteriorated insulation: Check for overheating evidenced by excessive discoloration of terminals and other visible copper; physical strains located by bent or distorted members.				
b. Terminals, including secondaries: Check for corrosion, loose connections.				
c. Secondary leads: Check for visible broken, cracked, or frayed insulation.				

E	U	R	D

d. Conduit and associated fittings carrying secondary leads: Check for rust, corrosion, other deterioration, loose joints in conduit fittings and around terminal boxes.

5. AUTOMATIC TAP-CHANGERS (load ratio control apparatus): Assess in accordance with manufacturer's instructions.
6. FORCED-AIR FANS AND FAN CONTROLS:
   a. Fans and motors: Check for defective bearings, inadequate lubrication, presence of dirt, bent or broken fan blades or guards, lack of rigidity of mountings, indications of corrosion or rust.
   b. Starting and stopping devices: Check for improper functioning as determined from operating once or twice.
   c. Fan speed: Check to ensure accordance with nameplate requirements.
7. WATER COOLING SYSTEMS: Water not being delivered in required quantity.
8. GAUGES AND ALARMS:
   a. Liquid level gauge and alarm system: Check for dirty, not readable, improper frequency of calibration.
   b. Pressure gauges and valves on inert gas systems: Check for improper frequency of gauge calibration; leaks in piping both before opening and after closing tanks; apply soap bubble test to all joints and connections when pressure is unsteady. Engineering tests should be performed under supervision of qualified electrical engineer before, during, or after inspection, as applicable. Assistance of craft personnel is required, and arrangements should be made with proper authority to ensure coordinated effort by everyone taking part.
   c. Test grounding system in accordance with Electrical Grounds and Grounding System Checkpoints.

E	U	R	D

d. Measure load current with recording meter over period of time when load is likely to be at its peak; measure peak-load voltage; make regulation tests and tests of operating temperature during peak-load-current tests; test and calibrate thermometers or other temperature alarm systems.

e. Test dielectric strength of insulating liquid, if necessary.

f. Test insulation resistance, if necessary.

Total Items: 8 Raw Total %

Findings (comment on each checkpoint and summarize to justify scoring):

____

____

____

____

____

____

____

____

____

____

____

E	U	R	D

Section 15. Power Transformers, Energized

SCOPE: Energized electric power transformers used for voltage transformation on transmission lines and high voltage distribution systems.

1. CONCRETE FOUNDATIONS AND SUPPORTING PADS:
   a. Check for settling and movement, surface cracks exceeding $\frac{1}{16}$ inch in width, breaking or crumbling within 2 inches of anchor bolts.
   b. Anchor bolts: check for loose or missing parts, corrosion, particularly at points closest to metal base plates and concrete foundations resulting from moisture or foreign matter, and exceeding $\frac{1}{8}$ inch in depth.
2. MOUNTING PLATFORMS, WOODEN:
   a. Check for cracks, breaks, signs of weakening around supporting members; rot, particularly at bolts and other fastenings, holes through which bolts pass, wood contacting metal.
   b. Check for burning and charring at contact points, indicating grounding deficiency.
   c. Check for inadequate wood preservation treatment.
3. MOUNTING PLATFORMS, METALLIC: Check for deep pits from rust, corrosion, other signs of deterioration likely to weaken structure.
4. HANGERS, BRACKETS, BRACES, AND CONNECTIONS: Check for rust, corrosion, bent, distorted, loose, missing, broken, split, other damage; burning or charring at wood contact points caused by grounding deficiency.
5. ENCLOSURES, CASES, AND ATTACHED APPURTENANCES:

E	U	R	D

a. Check for collection of dirt or other debris close to enclosure that may interfere with radiation of heat from transformer or flashover.
b. Check for dirt, particularly around insulators, bushings, or cable entrance boxes.
c. Check for leaks of liquid-filled transformers.
d. Check for deteriorated paint, scaling, rust; corrosion, particularly at all attached appurtenances, such as lifting lugs, bracket connections, and metallic parts in contact with each other.

6. NAMEPLATES AND WARNING SIGNS: Check for dirty, chipped, worn, corroded, illegible, improperly placed signs.
7. GASKETS: Check for leakage, cracks, breaks, brittleness.
8. INERT GAS SYSTEMS:
   a. Check for incorrect pressure in system (maximum: 3 to 5 pound; minimum: ¼ to 1 pound).
   b. Pipe and valve connections: Check for leaking gas (indicated by liquid oozing out of joints or by soapsuds test).
   c. Check for loose gas tank fastenings, loose valves.
   d. If previous arrangements were made with research forces, bleed a little gas from system and water gauge to ensure that fresh gas is let into system by means of pressure-regulating device; note evidence of leaks.
9. BUSHINGS AND INSULATORS: Check for cracked, chipped, or broken porcelain, indication of carbon deposits, streaks from flashovers, dirt, dust, grease, soot, or other foreign material in porcelain parts, signs of oil or moisture at point of insulator entrance.

	E	U	R	D
10. GROUNDING AND PHASE TERMINALS: Check for overheating evidenced by excessive discoloration of copper, loose connection bolts, defective cable insulation, no mechanical tension during temperature changes, leads appear improperly trained and create danger of flashovers from unsafe phase-to-phase or phase-to-ground clearances caused by deterioration of leads or expansions during temperature changes.				
11. INSTRUMENT TRANSFORMER JUNCTION BOXES AND CONDUITS: Check for loose or severely corroded components, including secondary lead connections.				
12. BREATHERS: Check for holes plugged with debris; desiccant-type breathers need servicing or replacement.				
13. TEMPERATURE-INDICATING AND ALARM SYSTEMS, INCLUDING CONDUIT AND FITTINGS: Check for loose fastenings, rust, severe corrosion, other mechanical defects, lack of lubrication, signs of burning around conducting and nonconducting parts of terminal boards.				
14. MANUAL AND AUTOMATIC TAP CHANGERS: Check for loose connections, rust, severe corrosion, other mechanical defects, lack of lubrication, signs of burning around conducting and nonconducting parts of terminal boards.				
15. LIQUID LEVEL INDICATORS: Check for rust, corrosion, lack of protective paint, cracked or dirty gauge glasses so that liquid level not discernible, plugged gauge-glass piping, liquid below permissible level indicated by mark for gauging, signs of leakage around piping, gauge cocks, gauge glasses, or other indicating devices.				
16. FANS AND FAN CONTROLS FOR AIR-COOLED TRANSFORMERS:				

E	U	R	D

   a. Check for lack of rigidity in mounting fastenings.
   b. Motors (external): Check for dirt, moisture, grease, oil, overheating, detrimental ambient conditions.
   c. Check for apparent deterioration of open wiring and conduit that may cause malfunctioning of either fans and controls.
   d. Check for improper functioning when manual (not automatic) fan controls operated.

17. WATER COOLING SYSTEMS:
   a. Check for leaks in piping, fittings, or valve; visible drainage system plugged; open ditches for drainage water fouled with vegetation.
   b. Bearings: Check for evidence of wear, indication of corrosion, external deterioration, leaks.
   c. Check for incipient deterioration, corrosion, rust, loose fastenings, other mechanical deficiencies, loose electrical connections for all components of alarm system.
   d. Temperature devices: Check for signs of deterioration that might cause malfunction or difficulty in taking readings.
   e. When pressure gauge readings on each side of strainer vary more than a pound or two, look for cause, such as plugged strainer.
18. GROUNDING: visual connections: Check for loose, missing, broken connections; signs of burning or overheating, corrosion, rust, frayed cable strands, more than one strand broken in 7-strand cable, more than 3 strands broken in 19-strand cable.
19. LIGHTNING ARRESTERS: Where attached to or mounted on, refer to Lighting Arresters Checkpoints.

E	U	R	D

20. PROTECTIVE RELAYS: Where attached to or mounted on, refer to Electric Relays Checkpoints.

Total Items: 20 Raw Total %

Findings (comment on each checkpoint and summarize to justify scoring):

____

____

____

____

____

____

____

Section 16. Safety Fencing

SCOPE: Metallic and wooden fencing around electrical power distribution equipment, including permanent barriers, intended to prevent unauthorized personnel from coming in contact with energized electrical equipment. Resistance of grounding system is covered by Electrical Grounding and Grounding Systems Checkpoints.

1. POST FOUNDATIONS AND EMBEDDED PIPE SLEEVES: Check for cracked, broken, settling, movement, water ponding at base; severe corrosion or need or recaulking of pipe sleeves.
2. METALLIC FENCE POSTS, GATE POSTS, HINGES, FASTENINGS RAILS, BRAC-

E	U	R	D

INGS, AND OTHER COMPONENT PARTS: Check for rust, corrosion, bent, loose, missing, inadequate, deteriorated paint.

3. GUARD WIRES AND GUARD WIRE BRACKETS: Check for rust, corrosion, bent, broken, loose, missing, sagging, failure to provide three guard wires where accessible to public (regardless of location), deteriorated paint.
4. WIRE FABRIC: Check for rust, corrosion, mechanical damage, holes 6 inches or more in diameter.
5. HOLDING WIRES, CLAMPS, AND OTHER FASTENINGS: Check for rust, corrosion, loose, missing, broken, other damage.
6. GATES: Check for rust, corrosion, bent, difficult operation, inadequate clearance, other damage; loose, missing, or other damage to stops, catches. Check latches, locks, and other appurtenances.
7. ENCLOSURE AND POSTS, RAILS, BRACES, GATES, AND OTHER WOODEN PARTS OF WOODEN FENCES AND WOODEN BARRIERS:
   a. Check for rot, loose, missing parts, leaning, broken, lack of rigidity, inadequacy, deteriorated paint.
   b. Supplemental guard wires: check for rust, corrosion. sagging, missing parts, broken, inadequacy of public safety, and prevention of entry of unauthorized persons.
8. ADEQUACY OF SECURITY THROUGHOUT ENTIRE PERIMETER:
   a. Check for construction excavations, washouts, unauthorized changes in ground surface grade both inside and outside fence line; openings, particularly along bottom edge, large enough to permit entry of average 5-year-old child.

E	U	R	D

b. Check for effective height decreased by changes of grade on either or both sides of fence.
c. Check for stacks of lumber, boxes, piles of dirt, or other materials stored within 3 feet of fence line on either side.
d. Check for possible installation of new equipment within 3 feet of fence line that will interfere with safety, maintenance, and inspection.

9. ADEQUACY FOR PREVENTION OF DAMAGE TO FENCE OR SAFETY OF PERSONNEL:
   a. Check for weeds, overgrowth, trash, or other debris along fence line.
   b. Check for damage to galvanizing or protective coatings from burning operations along fence line.
   c. Check for exposed live electrical parts less than 8 feet from inside of fence or barrier.
10. WARNING SIGNS:Not installed at all gates and other locations on fence, not in sight from each normal approach path, insecurely fastened, illegible.
11. ELECTRICAL GROUNDING:
    a. Cables not attached to post and fabric of fence; missing flexible connections at all gates.
    b. Check for rust, corrosion, frayed or broken, missing connectors, broken strands.
    c. Check for loose, missing, or other damage to visible connections at top of ground rods.
    d. Check for failure to connect or provide, or damage to ground wires attached to supplementary guard wires on wooden structures.

Total Items: 11 Raw Total %

E	U	R	D

Findings (comment on each checkpoint and summarize to justify scoring):

Section 17. Steel Poles and Structures

SCOPE: Steel poles and towers, metallic street-lighting standards, A-frames, and all other steel

E	U	R	D

structures used to support electric power lines or equipment, including those used for transmission lines, distribution systems, substations, and switching stations.

1. GROUND AREA: Trash and debris exceeding 2 quarts in volume, and weeds or brush 1 foot in height that are expected to continue growing, all within 3 feet of pole or structure.
2. CONCRETE BASES, PADS, AND ANCHOR BOLTS:
   a. Check for cracks, including surface cracks wider than $\frac{1}{16}$ inch, breaks, chipped areas deeper than $\frac{1}{2}$ inch, settlement, movement, water ponding at base.
   b. Check for rust and corrosion exceeding depth of $\frac{1}{16}$ inch, loose, missing, and bolt weaknesses, particularly at point where steel enters concrete.
3. STREET LIGHT STANDARD HANDHOLES AND BELL INTERIORS (visual inspection only, if energized)
   a. Check for rust, corrosion, or drops of moisture, indicating poor ventilation.
   b. Installed transformers: Check for loose wires, excessive discoloration from heating and sparking, signs of insulating compound or other leakage.
4. POLES, STRUCTURES, CROSSARMS, AND BEAMS (inspection from ground, use field glasses):
   a. Check for rust and corrosion exceeding 1 inch in diameter, in more than one or two spots, and closer than 2 or 3 feet; pits exceeding $\frac{1}{16}$ inch or larger spots where structural uprights enter concrete bases.
   b. Check for loose bolts and pins; excessive rust and corrosion between pole and/or structure and braces, equipment supports,

E	U	R	D

insulator pins, guy fastenings, and similar locations.

c. Check for checking, chipping, flaking, or scaling or paint on pole and attachments.
d. Check for broken or bent accessory members, especially near bolts.
e. Misalignment (top of unguyed pole is out of line more than 1 foot in any direction).

5. GUYS:
   a. Check for loose clamps, cracked or broken insulators, missing insulators where circuits exceed 300 volts.
   b. Check for broken strands, battered or rusty guy shields, indication of anchor looseness or movement.
6. GROUND WIRE (visual inspection only):
   a. Check for failure to install at least one wire at every steel pole or structural support, including each steel column in substation or switching station.
   b. Connections: Check for rust, corrosion, looseness, discoloration from overheating, substandard conditions.

Total Items: 6 Raw Total %

Findings (comment on each checkpoint and summarize to justify scoring):

____

____

____

____

____

E	U	R	D

Section 18. Vaults and Manholes (Electrical)

SCOPE: Vaults and manholes used in electrical power distribution system. Vaults include any space or structure used to house electrical distribution equipment. Manholes include in-ground structures used to provide junctions for cable runs, for pulling in cable, to allow space for expansion and contraction of cable, to provide ventilation, to drain underground conduit runs, or to house electrical equipment.

Transformers and switchgear are covered by other checkpoints.

1. MANHOLE COVERS AND GRATINGS:
   a. Check for dangerous, noxious, or flammable gases detected or suspected.
   b. Check for plugged vents, defective gaskets, cracks, rust, corrosion, particularly on underside, poor fit, structural inadequacy.

E	U	R	D

2. LADDERS AND STEPS: Check for rust, corrosion, loose anchorage, other defects.
3. ROOFS, WALLS, AND FLOORS:
   a. Check for dirt, evidence of burning, cracks, leakage, flooding, structural inadequacy, other defects.
4. VAULT DOORS: Check for unlocked, binding, difficult operation, does not swing clear and free; defective hinges, latches, locks, and other similar devices; rust, corrosion, abrasions, or other defects.
5. VENTILATING SYSTEMS, DUCTS, BLOWERS, AND AUTOMATIC CONTROLS: Check for dirt, rust, corrosion, excessive noise and vibration, dirty air filters, defective operation.
6. LIGHTS AND SWITCHES: Check for broken or missing globes and protectors; type not in accordance with safety regulations, if required; defective operation; rust and corrosion from excessive moisture.
7. FIREFIGHTING EQUIPMENT: Check for inadequate and apparent unsatisfactory operating condition (determine from fire inspector's tags).
8. SIGNS, INSTRUCTIONS, AND IDENTIFICATION TAGS: Check for dirt, illegible, and improperly located.
9. SEWER AND SUMP: Check for trash, other obstructions, clogged drains causing ponding or flooding, faulty operation of pump.
10. GROUNDING WIRE OR CABLE, AND GROUND RODS WHERE VISIBLE:
    a. Check for lack of continuity, loose connections, signs of corrosion.
    b. Measure ground-resistance values; report those in excess of 3 ohms.
11. BONDING SHEATHS:

E	U	R	D

a. Check for bonding wire or cable touching cable sheath at other than point of connection.
b. Check for lack of continuity; corrosion, loose connections.

12. CABLES:
    a. Check for excessive strain on sheath, poor arrangement, crowding, inadequate maintenance; splices for overheating, leaks, bulges, arrangement, splices.
    b. Check for cracks, punctures, deep scratches in cable sheath.
    c. Check for loose anchor bolts; missing cable racks, cable supports, broken or missing rack insulators, and defective or missing cable saddles.
13. DUCTS: Check for cable damage from abrasion, cable not free to expand and contract, inadequate cable training, defective or rusted cable shields.
14. FIREPROOFING CABLE (2200 volts or more) (visual inspection): Check for wrapping not in place, loose, insecure.
15. POTHEADS: Check for rust, corrosion, cracked or broken porcelain, leakage of joint compound.
16. SUBWAY JUNCTION BOXES:
    a. Check for rust, corrosion, loose, missing, or defective cover bolts.
    b. Check for breaks, aging, or leaking gaskets, particularly in pressurized boxes. If box is opened, check physical condition of sectionalizing fuses and copper links.

Total Items: 16 Raw Total %

Findings (comment on each checkpoint and summarize to justify scoring):

E	U	R	D

Section 19. Cathodic Protection Systems

SCOPE: Cathodic protection systems, including galvanic or sacrificial anode types and rectifier types installed for prevention and mitigation of corrosion of steel structures buried in the ground or in contact with water or other corrosive liquids. Electrical operation tests are not covered.

E	U	R	D

1. GALVANIC OR SACRIFICIAL ANODE SYSTEMS:
    a. Terminals and jumpers of test leads (permanently installed and accessible on underground systems): Check for rust, corrosion, broken or frayed wires, loose connections, similar deficiencies.
    b. Anode suspensions (elevated water tanks and systems for waterfront structures): Check for rust, corrosion, bent or broken suspension members or braces, frayed or broken suspension lines or cables, loose bolts, loose cable connections, frayed or broken wiring.
    c. Anodes (waterfront structures, where visible): when more than ¾ spent, report apparent average diameter remaining so arrangements can be made for replacement or adjustment.
    d. Bushing (supporting anode): Check for severe rust and corrosion; where resistors including variable types are installed in circuit, examine units for corrosion, broken or frayed wires, loose connections.
2. RECTIFIER-POWERED SYSTEMS:
    a. Exterior of enclosure: Check for rust, corrosion, mechanical damage.
    b. Cover hinges and locks: Check for rust, other deficiencies.
    c. Wiring and fastenings near rectifier: Check for broken or damaged insulation; rust, corrosion on conduit.
    d. Interior of enclosure: Check for rust, drops of moisture, loose wiring, signs of excessive heating. (Do not put hand or tools inside enclosure.)
    e. Record voltmeter reading, where installed.
    f. Record ammeter reading, where installed.
    g. Perform anode inspection.

E	U	R	D

h. Exposed wires and cables: Check for frayed or broken insulation.

i. Electrical connections (wires and cables connected to equipment except rectifier enclosures): Check for poor condition, loose connections, other deficiencies.

Total Items: 2 Raw Total %

Findings (comment on each checkpoint and summarize to justify scoring):

____

____

____

____

____

____

____

____

Section 20. Electric Motors and Generators

SCOPE: AC and DC electric motors, generators, exciters, motor-generator sets, synchronous converters and condensers, other electrical rotating equipment.

1. GENERAL: When practicable, start, run, and cycle motor and generator equipment through load range. Take care in starting motors and generators. On standby or infrequently oper-

E	U	R	D

ated equipment, check rotor freedom and lubrication. At humid locations, check records for evidence of regular exercise; if not found, arrange for drying out windows; megger windings before starting motor.

2. RUNNING INSPECTION (while equipment operates):
   a. Log or operator records: Check for evidence of motor or generator overload, induction motor underload, low power factor of load, excessive variations in bearing temperature, operating difficulties.
   b. Exposure: Check for unsafe accessibility for maintenance of instrumentation; exposed to physical or other damage from normal plant functions, processes, traffic, and radiant heat; inadequate personnel guards, fences; insufficient, missing, or illegible signs, identification, or operating instructions.
   c. Housekeeping: Check for dust, dirt, airborne grit, dripping oil, water, other fluids, vapors; corrosion; peeling, scratches, abrasion, or damage to painted surfaces.
   d. Machine operation: Check for noisy, unbalanced, excessive vibration, rattling parts.
   e. Structural supports: Check for inadequate, cracks, settlement; defective or inadequate vibration pads, shock mounts, dampers; loose, dirty, corroded bolts and fittings.
   f. Ventilation: Check for dirty, inadequate amount of air passing through machine; dirty, clogged, stator-iron air slots causing excessive temperature. (Too hot to touch. Measured temperature should not exceed 80°C for open frames or 90°C for enclosed frames. Compare with manufacturer's data.)
   g. Motor and generator leads: Check for exposed bare conductors; frayed, cracked,

E	U	R	D

peeled insulation; poor taping; moisture, paint, oil, grease; vibration, abrasions, breaks in insulation at entrance to conduit or machines; arcs, burns, overheated, inadequate terminal connections; lack of resiliency, lack of life, dried-out insulation; exposure to physical damage, traffic, water, heat, for semipermanent, temporary, or emergency connections.

h. Bearings: Check for improper lubrication (check lubrication schedules for lubricant used and frequency), improper oil level in oil gauges, incorrectly reading gauges, noisy bearings, overheated bearing caps or housing. (If bearings are too hot to touch, determine causes. A slow but continuous rise in bearing temperature after greasing indicates possible overlubrication or underlubrication, improper lubricant, or deteriorated bearings. Under normal conditions, the temperature of ball or roller bearings will vary from 10°F to 60°F above the ambient temperature.)

i. Collector Rings, Commutators, Brushes: Check for excessive sparking, surface dirt, grease (check cleanliness with clean canvas paddle); sparking or excessive brush movement caused by eccentricity, sprung shaft, worn bearings, high bars or mica, surface scratches, roughness; end play resulting from magnetic-center hunting of rotor; inadequate brush freedom; nonuniform brush wear; poor commutation, improper brushes, incorrect brush pressure. Adjust brush spring pressure to between $1\frac{3}{4}$ to $2\frac{1}{2}$ psi of brush–commutator contact area for light metallized, carbon or graphite brushes; for pressure for other type brushes, check manufacturer's data. (Measure with spring scale.)

E	U	R	D

j. Starters, motor controllers, rheostats, associated switches: Check for damaged or defective insulation, laminations, defective heater or resistance elements, worn contacts, shorts between contacts, arcing, grounds, loose connections, burned or corroded contacts.

k. Protective equipment: Check for dirt, signs of arcing, symptoms of faulty operation, improper condition of contacts, burnt-out pilot lamps, burnt-out fuses.

3. SHUTDOWN INSPECTION (while equipment is not in operation and is electrically disconnected; a shutdown inspection includes a running inspection):
   a. Stators: dirt, debris, grease; coils not firmly set in slots; burns, tears, aging, embrittlement, moisture in insulation; clogged air slots; rubbing, corrosion, loose laminations of stator-iron; charred or broken slot wedges; abrasion of insulation or chafing in slots; signs of arcing or grounds.
   b. Rotors: check for difficult turning, rubbing, excessive bearing friction, end play, overheating, looseness of windings, charred wedges, broken, cracked, loosely welded or soldered rotor bars or joints; cracked end rings in squirrel cage motors; loose field spools and deteriorated leads and connections in synchronous motors; deteriorated insulation in rotors.
   c. Rotor-stator gaps: Check gaps on 5-hp or larger induction motors, particularly of the sleeve bearing type. Where practicable, measure and record gaps on the load, pulley, or gear end of the motor. Measure at two rotor positions, 180° apart, 4 points for each rotor position. If there is more than 10% variation in gaps, arrange for realignment.

E	U	R	D

d. Mechanical parts: Check for corrosion, improper lubrication, misalignment, end play, interference, inadequate chain or belt tension.

e. Insulation resistance: Test insulation resistance of motor and generator windings. Compare results with Table 2.4. Insulation resistance values are arbitrary and should be correlated with operating conditions, exposure to moisture, metallic dust, age, length of time in service, severity of service, and maintenance levels.

f. Verify if permanent records are kept of measure insulation values on all integral hp motors. To ascertain trends, graphs showing the long-term relationship of insulation resistance to time should be prepared on all large or critical equipment.

Total Items: 3 Raw Total %

Findings (comment on each checkpoint and summarize to justify scoring):

____

____

____

____

____

____

____

____

TABLE 2.4. Insulation-Resistance Values

Machine Rating (volts)	Insulation Resistance (megohms)	
	Minimum	Preferred
110	0.11	0.20
220	0.22	0.50
440	0.42	0.75

Inspection Trouble Chart—Probable Source of Trouble

Evidence/Probable Cause	DC Motors		DC Generators	AC Motors			AC (Synchronous) Generators
	Shunt	Series		Squirrel Cage	Slip Ring	Synchronous	
Hot bearings:							
Improper-grade lubricant	x	x	x	x	x	x	x
Lubricant w/excessive moisture	x	x	x	x	x	x	x
Overlubrication or under lubrication	x	x	x	x	x	x	x
Dirty lubricant	x	x	x	x	x	x	x
Loss of lubricant	x	x	x	x	x	x	x
Clogged oil grooves or vents	x	x	x	x	x	x	x
Broken or excessively worn oil rings	x	x	x	x	x	x	x
Oil rings stuck or out of slots	x	x	x	x	x	x	x
Tight bearing caps (inadequate clearance)	x	x	x	x	x	x	x
Loose bearing	x	x	x	x	x	x	x
Poor alignment (foundation settlement)	x	x	x	x	x	x	x
NOISY BEARINGS:							
(Rumbling or growling): Rough races (brinelled)	x	x	x	x	x	x	x
Corroded balls, rollers or races	x	x	x	x	x	x	x
Excessive clearance (wear, corrosion, flattened or cracked balls, rollers, or races)	x	x	x	x	x	x	x

Inspection Trouble Chart—Probable Source of Trouble (*Continued*)

Evidence/Probable Cause	DC Motors		DC Generators	AC Motors			AC (Synchronous) Generators
	Shunt	Series		Squirrel Cage	Slip Ring	Synchronous	
NOISY BEARINGS: (popping or churning): Overlubrication	x	x	x	x	x	x	x
Excessive moisture in lubricant	x	x	x	x	x	x	x
EXCESSIVE ARCING AT BRUSHES:							
Incorrect brush position	x	x	x		x	x	x
Improper type, size, or span of brushes	x	x	x		x	x	x
Incorrect brush pressure or contact	x	x	x		x	x	x
Loose brush rigging	x	x	x		x	x	x
Dirty or rough commutator or slip ring	x	x	x		x	x	x
High or low commutator bars	x	x	x				
Short-cicuited commutator bars	x	x	x				
Overload or excessive vibration	x	x	x				
NOISY BRUSHES (SINGING):							
Excessive brush pressure	x	x	x		x	x	x
Brushes too hard	x	x	x		x	x	x
Holders improperly adjusted	x	x	x		x	x	x
NOISY BRUSHES (CHATTERING):							
High or low commutator bars	x	x	x				
Loose commutator bars	x	x	x				
High mica between cummutator bars	x	x	x				
Brushes set at improper angle	x	x	x		x	x	x
RING FIRE AND FLASHING ON COMMUTATOR:							
Short or open circuit in armature coil	x	x	x				

OVERHEATING:							
Overload	x	x	x	x	x	x	x
Field or armature short circuit	x	x	x	x	x	x	x
Poor ventilation	x	x	x	x	x	x	x
Rotor off center	x	x	x	x	x	x	x
Unbalanced phase current						x	x
Excessive field current						x	x
Line voltage too low	x	x	x	x	x	x	x
Bearing friction	x	x	x	x	x	x	x
SPEED TOO HIGH:							
Weak field current	x	x					
Prime mover speed too high	x	x					
SPEED TOO LOW Overload	x	x		x	x		
Low line voltage	x	x		x	x		
Bearing friction	x	x		x	x		
Dragging rotor	x	x		x	x		
Prime mover speed too low			x			x	
HUNTING ACTION:							
Load variation	x	x					
Variation in voltage frequency (unstable speed of prime mover)			x	x	x		

E	U	R	D

Section 21. Pier Circuits and Receptacles

SCOPE: These checkpoints cover electrical circuits for light and power on piers. Included are the cable and wire circuits from the distribution center or substation to the pier service outlets. Utilization equipment and portable cables are not covered. Vaults and manholes are covered by Vaults and Manholes (Electrical) Checkpoints.

1. CABLE AND WIRE:
   a. Exposed runs: unsafe, unreliable, evidence of overheating, grounds, short circuits, improper and unauthorized connections, bare conductors hazardous to personnel, damaged or defective insulation, exposure to physical damage, defective wiring devices and accessory fittings, improper terminations.
   b. Exposure to moisture, greases, oils, heating; excessive sag and vibration; poor ventilation; crowded ducts or conduit; improper spacing and use of duct, conduit, cable, and wire.
   c. Broken or missing parts, leaking oil filler compounds, overheated joints, terminations, and splices.
2. INSULATION IN ALL CIRCUITS, SERVICE AND SWITCHING HOODS, CIRCUIT TERMINATIONS, PANELS, SHIP-TO-SHORE OUTLETS, AND ALL VISIBLE LOCATIONS:

E	U	R	D

Check for physically deteriorated insulation, evidence of overheating or burning, dirty, exposure

3. WEATHERPROOF ENCLOSURES, INCLUDING BOXES, CABINETS, AND HOODS THAT HOUSE CABLE TERMINATIONS, SWITCHES, PIER OUTLET BREAKERS, AND SERVICE OUTLETS:
   a. Check for dirt and dust, rust, leakage, clogged enclosure drain; chafing, moisture, improper entrance, inadequate terminations of conduit or cable at entrance.
   b. Check for unserviceable covers and seals; inoperable or inadequate hinges, locks, and handles; rusted, corroded, loose, missing nuts, bolts, and screws.
   c. Check for inadequate or illegible identification labels, other markings; evidence of lines, cables, hawsers being tied to enclosures.
4. RECEPTACLES, OUTLETS, AND CONDUITS:
   a. No protection against dirt, weather, and entrance of moisture; dirty contacts, evidence of flashovers or overheating, improper grounding, inadequate or illegible identifications, loose fittings, other conduit deficiencies.
   b. Threaded caps: cross-threaded, broken chains, loose connections, defective seals.
   c. Swing-type caps or open type with rubber flaps: poor closure and sealing, allowing entrance of dirt and moisture.
5. PANELS:
   a. Not serviceable, dirty, rust, corrosion, loose connections, unprotected countersunk bolts.
   b. Ground connections: lack of continuity, loose.

E	U	R	D

c. Illegible and inadequate written instructions, including phase and polarity markings.

6. PIER OUTLET BREAKERS AND SWITCHES:
   a. Improper functioning, dirty, contact misalignment, inadequate and loose connections.
   b. Rust and corrosion on metal parts; defective gaskets for covers and hoods.
   c. Improper functioning of indicating lamps; inadequate heaters for condensation prevention.
7. PIER LIGHTING LUMINARIES, STANDARDS, FLOODLIGHTS, SWITCHES, AND ALL METAL PARTS: Dirty, rust, corrosion, loose, missing, broken, defective operation.
8. TESTS:
   a. Measure insulation resistance between cable conductors and ground in pier circuit feeders; keep running record of insulation resistance measurements, noting test points, instrument used, and dates of test; report evaluations of such measurements; insulation resistance measurements should exceed 300,000 ohms for safe operation.
   b. Spot-check illumination levels; light output depreciation of 20 to 25% below level obtainable from clean fixtures and new lamps will require lamp replacements.

Total Items: 8 Raw Total %

Findings (comment on each checkpoint and summarize to justify scoring):

____

E	U	R	D

Section 22. Distribution Transformers, Deenergized

SCOPE: Deenergized electric distribution transformers used for voltage reduction. Before inspection, make arrangements to have electricians and other required labor available.

1. BUSHINGS AND INSULATORS:
   a. Insulators and porcelain parts: indications of cracks, checks, chips, breaks; where flashover streaks are visible, reexamine for injury to glaze or for presence of cracks.
   b. Report for further investigation any chipped glaze exceeding ½ inch in depth or an area exceeding 1 square inch on any insulator unit.

E	U	R	D

c. Severe cracks, chipped cement, or indications of leakage around bases of joints of metal to porcelain parts at terminal and transformer ends.
d. Terminal ends: mechanical deficiencies, looseness, corrosion, damage to cable clamps.
e. Improper oil level in oil-filled bushings.
f. Heating evidenced by discoloration, and corrosion indicated by blue, green, white, or brown corrosion products on metallic portions of all main and ground terminals, including terminal board and grounding connections inside transformer case.

2. ENCLOSURE AND CASES: If case is opened for any reason, examine interior immediately for signs of moisture inside cover and, where present, for plugged breathers, inactive desiccant, enclosure leakage.
3. COILS AND CORES: When cover is open, examine interior for deficiencies, dirt, sludge. If feasible, probe down sides with glass rod, and if dirt and sludge exceed approximately ½ inch, report for change or filter insulating oil and have coils and cores cleaned.
4. GAUGES AND ALARMS:
   a. Liquid level gauge and alarm system: dirty, not readable, improper frequency of calibration. Engineering tests should be performed under supervision of qualified engineer before, during, or after inspection, as applicable. Assistant inspectors and craft personnel are required, and arrangements should be made with proper authority to ensure coordinated effort by everyone taking part.
   b. Test grounding system in accordance with Electrical Grounds and Grounding Systems Checkpoints.

E	U	R	D

c. Measure load current with recording meter over period of time when load is likely to be at its peak; measure peak-load voltage; make regulation test and tests of operating temperature during peak-load-current tests; test and calibrate thermometers or other temperature alarm systems.
d. Test dielectric strength of insulating liquid.
e. Test insulation resistance.

Total Items: 4 Raw Total %

Findings comment on each checkpoint and summarize to justify scoring):

____

____

____

____

____

____

____

____

____

____

____

____

E	U	R	D

Section 23. Disribution Transformers, Energized

SCOPE: Energized electric distribution transformers used for voltage reduction.

1. CONCRETE FOUNDATIONS AND SUPPORTING PADS:
   a. Settling and movement, surface cracks exceed $1 \frac{1}{16}$ inch in width, breaking or crumbling within 2 inches of anchor bolts.
   b. Anchor bolts: loose or missing parts, particularly at points closest to metal and concrete foundations resulting from foreign matter, and exceeding $\frac{1}{8}$ inch
2. MOUNTING PLATFORMS, WOODEN:
   a. Check for cracks, breaks, signs of weakening around supporting members; rot, particularly at bolts and other fastenings, holes through which bolts pass, wood contacting metal.
   b. Check for burning and charring at contact points, indicating grounding deficiency.
   c. Check for inadequate wood preservation treatment.
3. MOUNTING PLATFORMS, METALLIC: Check for deep pits from rust, corrosion, other signs of deterioration likely to weaken structure.
4. HANGERS, BRACKETS, BRACES, AND CONNECTIONS: Check for rust, corrosion, bent, distorted, loose, missing, broken, split, other damage; burning or charring at wood contact points resulting from grounding deficiency.

E	U	R	D

5. ENCLOSURES, CASES, AND ATTACHED APPURTENANCES:
   a. Check for collection of dirt or other debris close to enclosure that may interfere with transfer of heat from transformer or flashover.
   b. Check for dirt, particularly around insulators, bushings, or cable entrance boxes.
   c. Check for leaks of liquid-filled transformers.
   d. Check for deteriorated paint, scaling, rust; corrosion, particularly at all attached appurtenances, such as lifting lugs, bracket connections, and metallic parts in contact with each other.
6. NAMEPLATES AND WARNING SIGNS: Check for dirty, chipped, improperly placed.
7. GROUNDING: Visual connections: check for loose, missing, broken connections; signs of burning or overheating, corrosion, rust, frayed cable strands, more than 1 strand broken in 7-strand cable, more than 3 strands broken in 19-strand cable.
8. BUSHINGS AND INSULATORS: Check for cracked, chipped, or broken porcelain, indication of carbon deposits, streaks from flashovers, dirt, dust, grease, soot, or other foreign material on porcelain parts, signs of oil or moisture at point of insulator entrance.
9. GROUNDING AND PHASE TERMINALS: Check for overheating evidenced by excessive discoloration of copper, loose connection bolts, defective cable insulation, no mechanical tension during temperature changes, leads appear improperly trained and create danger of flashovers from unsafe phase-to-phase or phase-to-ground clearances caused by deterioration of leads or expansions during temperature changes.

E	U	R	D

10. LIGHTNING ARRESTERS: Where attached to or mounted on, see Lightning Arresters Checkpoints.
11. BREATHERS: Check for holes plugged with debris; desiccant-type breathers need servicing or replacement.
12. GRILLS AND LOUVERS FOR VENTILATION OF AIR-COOLED, TRANSFORMERS: Check for plugged with debris or foreign matter, interfering with free passage of air. (Openings located near floor or ground line can be inspected with small nonmetallic framed mirror having long insulated handle, used in conjunction with light from hand flashlamp having insulated casing. Throw light beam onto mirrors and reflect upward into openings.)

Total Items: 12 Raw Total %

Findings (comment on each checkpoint and summarize to justify scoring):

___

___

___

___

___

___

___

___

E	U	R	D

Section 24. Buried and Underground Telephone Cable

SCOPE: Buried and underground lead-covered telephone cables, including cable trench and trench markers.

1. BURIED CABLE (walk over all areas where buried cable is known to exist):
   a. Check for evidence of construction work interference or damage.
   b. Check for sunken trench, indicating depressions so pronounced that a drop in trench base is evident. (Have test excavation made to disclose deficiencies.)
   c. Check for trench markers displaced, missing, or damaged.
   d. Check that men at work instructed not to damage cable and their supervisor instructed to station a person to see that warning is carried out.
2. UNDERGROUND CABLE:
   a. Manholes: check for loose, poor fit, or missing covers; flooding, excessive moisture, seepage of water through walls or floor and around duct entrances.
   b. Check for loose duct plugs allowing water or gas seepage.
   c. Cable racks and ties: Check for looseness, corrosion.

E	U	R	D

d. Cables: check for evidence of corrosion; area of fine cracks and granular appearance on sheath metal, indicating crystallization, particularly near point where cable leaves duct and at sharp bends.
e. Check for electrolytic action on cable sheath, sleeves, and threatening wiped joints.
f. Drainage wires: Check for defective fuses, poor connector to cable and buss bar or racks.
g. Test for maximum and minimum values of current flowing in drainage wire and note appreciable differences in amounts measured compared with measurement of previous tests.

Total Items: 2 Raw Total %

Findings (comment on each checkpoint and summarize to justify scoring):

____

E	U	R	D

Section 25. Telephone Substations

SCOPE: Telephone substations, their wiring, and associated equipment.

1. TERMINAL: Check for defective face plate, broken lugs, dirt; connections for crosses and shorts; drop wire attachments and connections for looseness.
2. DROP OR BLOCK WIRE FROM TERMINAL (visual from ground): Check for deteriorated, damaged, improperly placed, inadequately supported, insufficient clearances from trees and utilities, damaged insulation, missing or damaged guards.
3. PROTECTOR:
   a. Check if improperly located with respect to liability to damage from moisture or mechanical injury.
   b. Check for loose, inside wire connections. Test ground connections for ground at protector.
4. INSIDE WIRE (where accessible):
   a. Check if improperly or not securely fastened, damaged or defective insulation, inadequate clearance or insulation from electric wire, water pipes, etc.
   b. Check if existing location subjects wire to liability of damage from moisture, mechanical injury, or other cause.
   c. Connection block, if used: check for improper mounting, loose connections, bent cover in contact with terminals.

E	U	R	D

d. Ground Connection: Check if defective, improperly made, subject to mechanical injury, insufficient separation from light or power wires is evident.

5. TRANSMITTER: Mouthpiece: cracks, breaks.
6. RECEIVER:
   a. Shell: cracks, breaks.
   b. Diaphragm: dents, rust, etc.
7. CASE: Dirty, cracked, broken, rubber or felt pads missing on desk sets.
8. DIAL: Insecurely mounted, bent wheel or finger stop. Test speed and functioning of dial.
9. CORDS: Frayed or worn; secure loose stay cords and hooks; straighten twisted cords.
10. CAPACITORS: Leads improperly soldered, terminals loose.
11. RINGER: Improperly adjusted; insecurely mounted; loose gongs; nicks, dents, or improper connections in coils.
12. INDUCTION COIL: Insecurely mounted; nicks, dents, or breaks in windings.
13. HOOK SWITCH: Tight lever action; bends, rust, pitting, springs; contacts make and break improperly.
14. WIRE FORMS: Incorrect color code and connections, broken or frayed wires, poor soldered connections, defective lacings, inadequate wire lengths.
15. SCREWS AND NUTS: Missing, stripped, or worn screws and nuts.

Total Items: 15 Raw Total %

Findings (comment on each checkpoint and summarize to justify scoring):

E	U	R	D

Section 26. Fuses and Small Circuit Breakers (600 Volts and Below, 30 Amperes and Below)

SCOPE: Visual inspection only of fuses and small circuit breakers and their enclosures in electrical circuits operating at 600 volts or below and rated at 30 amperes and below.

1. BYPASSING: Report apparent bypassing of fuses or circuit breakers for further investigation.
2. HOUSEKEEPING: Check for dust, dirt, oil, grease, corrosion, foreign matter within enclosure; inadequate identification of circuits.
3. ENCLOSURES: Check for deterioration of enclosures connecting conduit or cable due to rust corrosion; loose, corroded, or missing covers.

E	U	R	D

4. CONNECTIONS (if visible without removing covers): Check for loose, corroded, inadequate; deteriorated insulation.
5. CAPACITY: Check size of existing fuses or circuit breakers against system engineering drawings. Report oversized fuses and circuit breakers.
6. FUSES: Check for overheating, indicated by discoloration of brass or copper at contact points; distortion, charring, deterioration of fiber cases of cartridge-type cases.
7. CIRCUIT BREAKERS: Check for distortion, charring, deterioration of molded portions of case.
8. GROUNDING: Check for loose, corroded connections; deteriorated or abraded insulation; frayed or broken cables.

Total Items: 8 Raw Total %

Findings (comment on each checkpoint and summarize to justify scoring):

____

____

____

____

____

____

____

____

E	U	R	D

Section 27. Rectifiers

SCOPE: Visual inspection only of metallic and mercury-arc rectifiers. Rectifiers used in cathodic protection systems are covered in Cathodic Protection Systems Checkpoints.

1. ENCLOSURES:
   a. Housekeeping: check for dust, dirt, trash, debris in general area.
   b. Exterior: check for mechanical damage, excessive corrosion (more than two rust spots ½ inch diameter); corroded, binding, unlubricated hinges and latches.
   c. Interior: check for rust, corrosion, moisture, condensation, indication of excessive heating.
   d. Wiring: check for broken, damaged, deteriorated, missing insulation or clamps; corroded or mechanically damaged conduit; cracked or broken sleeves on floor or wall bushings.
2. METALLIC RECTIFIERS:
   a. Electric meters: record readings from all AC and DC ammeters and voltmeters. Report if supply voltage is more than 5% below or above nameplate rating.

E	U	R	D

b. Temperature: record readings of water temperature indicators, if provided. On indoor installation, record ambient temperature at apparent hottest point 5 feet from units. Report if temperature of cooling water is more than 10% above that recommended by manufacturer.

c. Fan: Check for dirt, excessive vibration, bolts, loose hold-down bolts, loose or worn bearings, improper lubrication.

3. MERCURY-ARC RECTIFIERS:

   a. Meters and gauges: Check for illegible; inadequate lighting; cracked, broken, dirty, badly stained viewport glasses.

   b. Water cooling systems: check for leaks, rust, corrosion mechanical damage, excessive vibration.

   c. Pumps, fans, and motors: Check for leaks, excessive vibration, loose or missing hold-down bolts, deteriorated mounting pads or shock pads, inadequate or improper lubrication.

Total Items: 3 Raw Total %

Findings (comment on each checkpoint and summarize to justify scoring):

____

____

____

____

E	U	R	D

Utilities and Ground Improvement Assessment

E	U	R	D

I. *OBJECTIVE:* The basic objective is to maintain utilities and ground improvements in an economical manner that will protect the organization's investment, reduce hazards to life and property, and permit continued service consistent with the mission.

II. *DEFINITIONS:*

A. Utilities and grounds improvements are classified as follows:

1. Bridges, Trestles
2. Fences, Walls, Gates
3. Grounds—improved, semi-improved, unimproved
4. Railroad trackage
5. Pavement—roads, streets, sidewalks, airfield, aprons, parking areas, curbs, airfield runways, open storage, embankments, ditches, culverts
6. Retaining walls
7. Storm drainage systems
8. Tunnels and underground structures piping

E	U	R	D

9. Piping system
10. Steam distribution equipment
11. Fresh water supply and distribution systems
12. Pumps (sump, sewage, water)
13. Sewage collection and disposal systems
14. Unfired pressure vessels
15. Underground tanks

B. *SECTIONS:* Sections for which inspections checklists are established are depicted above.

III. *MAINTENANCE STANDARDS:* The degree of maintenance, repair, and rehabilitation of utilities and grounds improvements shall be governed by known foreseeable usage. All items in this category shall be maintained to the extent necessary to ensure reliability of service, full safety of operations, efficient operation of the equipment/items, and prevention of unwarranted deterioration.

Section 1. Bridges and Trestles

SCOPE: Bridges and trestles, including those constructed of steel, timber, masonry, concrete, and composite materials. Not covered are concrete boxes with integral floor, which are classified as culverts regardless of span. These are included under drainage structures.

1. *SIDE SLOPES:* Check for failure to maintain slopes of 1 1/2 to 1 or more; soil erosion; inadequately protected with vegetation or mulch; concrete overlays (if applicable); cracking, spalling, broken areas, other damage.
2. *BRIDGE AND FOUNDATION PROTECTIVE STRUCTURES, SUCH AS RIPRAP*

E	U	R	D

CRIBBING, BULKHEADS, DOLPHINS, PILES, OR OTHER: Check for missing, broken, insect and other pest infestation, decay, erosion, undermining, scouring, other damage.

3. *DRAINAGE DITCHES:* Check for loose bottom and sides; improper side sloping; silting; failure to protect surrounding areas at outfalls from erosion.
4. *ROADWAY OF APPROACHES:* Check for cracked, broken, alligatored, and disintegrated concrete or bituminous surfaces: cracked, broken, or other damage to curb and gutter sections.
5. *APPROACH FILL:* Check for settlement, particularly at joint between fill and structure.
6. *FENCES, BARRICADES, AND RAILINGS AT APPROACHES:* Check for inadequacy or structural damage; missing or illegible load and speed limits.
7. *DRAINAGE CHANNELS:* Check for erosion, scouring, accumulations of driftwood and debris above, below, and at structure; evidence of possible course diversion resulting from obstructions, erosion, or other.
8. *CONCRETE FOUNDATIONS:* Check for cracks, scaling, disintegration, exposed reinforcing. Wood piling and pads: missing, broken, ineffective bearing, decay, termite and other pest infestation. All foundations: scouring, undermining, settlement.
9. *ABUTMENTS AND PIERS:* Check for cracks, breaks, scaling, spalling, disintegration, open joints, other damage; evidence of damage from impact and vibration; failure of expansion devices; damage from floating debris, ice, and waterborne traffic.
10. *TIMBER FRAMING:* Check for loose, missing, twisted, bowed, warped, split, checked,

	E	U	R	D
unsound members; deteriorated joints; rot, termite and other insect infestation.				
11. *STEEL, FRAMING:* Check for rust, corrosion; loose, missing, bowed, bent, broken members.				
12. *CONCRETE AND MASONRY STRUCTURES:* Check for weathering, cracks, spalling, exposed reinforcing; open, eroded, or sandy mortar joints; broken and missing stones.				
13. *ALL SUPERSTRUCTURES:* Check for damage from floating debris, ice, and waterborne traffic; misalignment both horizontal and vertical.				
14. *WOOD FLOORING:* Check for loose, missing, broken, rotted pieces; protruding nails and other fastenings; checkered wearing plates; loose, missing, or other damage.				
15. *STRUCTURE ROADWAYS:* Check for cracked, broken, corrugated, disintegrated concrete or bituminous surfaces.				
16. *CONCRETE CURBS AND GUTTERS AND / OR CONCRETE OR MASONRY HANDRAILS AND HANDRAIL WALLS:* Check for loose, missing, and broken individual sections; misalignment; sandy and eroded mortar joints; loose or missing capstones; other damage.				
17. *EXPANSION JOINTS:* Check for improper sealing; loose or missing filler; failure to allow movement when filled with trash or debris.				
18. *METAL HANDRAILS:* Check for rust, corrosion, loose, missing, broken, misalignment, other damage.				
19. *BRIDGE SEATS, BEARING AND COVER PLATES:* Check for rust, corrosion, missing, loose, other damage.				

	E	U	R	D
20. *ROLLERS AND OTHER SIMILAR DEVICES:* Check for rust, corrosion, inadequate lubrication, failure to allow movement.				
21. *CABLES:* Check for frayed, raveled, or broken strands; inadequate lubrication; defective anchorage; interference from overhanging objects.				
22. *SPLICES, BOLTS, RIVETS, SCREWS, AND OTHER CONNECTIONS:* Check for rust, corrosion, loose, missing, broken welds, other damage.				
23. *MOVABLE BRIDGES:* Check for rust, corrosion, wear, inadequate lubrication. (Examine through complete operating cycle, if required.)				
24. *UTILITY SUPPORTS:* Check for rust, corrosion, loose, missing, or broken parts.				
25. *UTILITY LINES:* Check for corrosion, leaks, sagging, insulation and waterproofing defects, mechanical damage.				
26. *PAINTED SURFACES:* Check for rust, corrosion, cracking, scaling, peeling, wrinkling, alligatoring, chalking, fading, complete loss of paint.				
Total Items: 26 Raw Total %				

Findings (comment on each checkpoint and summarize to justify scoring):

____

____

____

____

____

E	U	R	D

Section 2. Fences and Walls

SCOPE: Security fences and all other outside partition fences and walls, except for safety fencing around electrical power distribution substations, and retaining walls. These expected items are covered by separate checkpoints.

1. *FENCES:* Check for discontinuity, looseness, vertical and horizontal misalignment, erosion that would permit entry of unauthorized persons or animals.
2. *FABRIC:* Check for looseness, rust, corrosion, broken areas, holes, and loose, missing, broken, or other damage to guard and stretch wires, and fastening wires and clamps; particularly at endposts, cornerposts, gateposts, and where attached to a structure.
3. *METAL POSTS:* Check for rust, corrosion; loose, bent, leaning, broken, or missing or mechanically damaged posts; inadequate

E	U	R	D

base support; settlement of concrete foundation.

4. *TOP, BOTTOM AND METAL MID AND BRACING RAILS:* Check for rust, corrosion, bent, broken, or missing, particularly at corners.
5. *METAL GATES:* Check for misalignment, difficult opening and closing, and loose, missing, or broken stops, checks, rollers, hinges, latches, and locks.
6. *WOOD PICKETS OR PLANKS, AND RAILS:* Check for loose, missing, broken, decay, and insect infestation; failure to provide ground clearance under fence pickets, planks, or bottom rails.
7. *WOOD POSTS:* Check for loose, leaning, splintered, broken, missing, or mechanically damaged posts; rot, termite infestation; improper wood species or failure to receive treatment to resist damage from weathering or insects.
8. *WOOD GATES:* Check for loose, broken, splintered, rotted, or missing parts, misalignment; difficult opening and closing; loose, missing, or broken stops, checks, rollers, hinges, latches, and locks.
9. *SUPPLEMENTARY WIRE GUARDS AND ATTACHMENTS:* Check for rust, corrosion, loose, missing, broken, or other damage.
10. Check for presence of weeds, trash, or other debris along fence line and growing on fence and damage to metal or wood parts from burning operations. Trees or shrubs should not be allowed to grow close to security fences. Vegetation should not be allowed to grow more than 8 inches high along fences and walls. Vines should not be allowed to grow on fences.

E	U	R	D

11. *PAINTED SURFACES:* Check for rust, corrosion, flaking, scaling, peeling, blistering, or complete absence of paint.
12. *CONCRETE AND MASONRY SURFACES:* Check for cracks, spalling, broken areas, settlement, eroded and sandy mortar joints.
13. Check for ponding water or soil erosion at foundations.
14. Check for loose, missing, or broken capstones.
15. *WOOD WALLS:* Check for broken, rotted, termite or other insect infestation; complete deterioration, out of plumb, discontinuity.

Total Items: 15 Raw Total %

Findings (comment on each checkpoint and summarize to justify scoring):

____

____

____

____

____

____

____

____

____

____

E	U	R	D

Section 3. Grounds

SCOPE: Lawn and turf areas, areas seeded to rough grasses, agricultural and grazing lands, woodlands, trees, and shrubs, runoff and erosion control works, fill and cut slopes, gullies, irrigation systems, and weed control.

1. *LAWN AND OTHER TURF AREAS INCLUDING BORDERS:* Check for traffic damage; loss of color; density; sparse and bare spots; weeds; undesirable grasses; diseases; insect damage; erosion; silt deposits; waterborne debris; excessive height; bruised or damaged ends from dull mower.
2. *TREES AND SHRUBS IN LANDSCAPED AREAS:* Check for lack of vigor, need of trimming, interference with utilities or buildings, injury from mowers, structural weaknesses, and storm, disease, or insect damage.
3. *BORDER STRIPS AND AREAS SEEDED TO ROUGH GRASSES FOR EROSION CONTROL:* Check for poisonous or noxious weeds; seedling trees that may hinder future mowing; erosion and siltation; lack of vigor; inadequacy of coverage; evidence of burning.
4. *WOODLANDS:* Check for erosion; dead, diseased, or damaged trees; firelanes for impassability; vegetation growth that may carry ground fires; hollow trees.
5. *EARTH DAMS AND DIKES:* Check for damage from erosion, burrowing animals; seepage; lack of vegetation density or vigor of growth; drop inlet pipes for stoppage; logs,

E	U	R	D

debris; outlet ends for erosion, seepage, piping damage or failure.

6. *EMERGENCY SPILLWAYS OF DROP INLET DAMS:* Check for blockage, erosion damage.
7. *PERMANENT CHECK DAMS IN WATER COURSE:* Check for overflow at notch section, bypassing at ends, erosion on downstream side, damaged and deteriorated walls and apron.
8. *HILLSIDE AND TERRACE DIVISION EMBANKMENT, CHANNELS, AND CULVERTS:* Check for silt, debris, rank vegetation, low and weak sections, overflow, erosion, gullying, burrowing animals.
9. *VALLEY DRAINAGE CHANNELS, INCLUDING CULVERTS AND LATERAL DRAINS AND TILE AT ENTRANCE POINTS:* Check for overflow, stoppage, silt, debris, rank vegetation, erosion, caving, sloughing scour.
10. *VEGETATED WATERWAYS:* Check for inadequate vegetation fullness and cover in relation to ground surface area that should be shielded; erosion of waterway and along sides; debris, overflow.
11. FILL SLOPES ON BARRICADES, HIGHWAYS, RAILWAYS, AIRFIELD RUNWAYS, IGLOOS, AND OTHER SOIL-COVERED BUILDINGS: Check for erosion, burning, steepness; lack of vigor and insufficient vegetation coverage for protection against beating rain and direct sunshine; inadequate fill depth at top of slope wherever buildings and weather conditions necessitate variations on different slopes; inadequate surface runoff piping; insufficient thickness of inorganic mulch (gravel, slag, etc.).
12. *CUT SLOPES AND DIVERSION CHANNELS:* Check for erosion, scour, burning,

E	U	R	D

weaknesses from past or possible overflow, lack of vigor or growth and insufficient vegetation coverage; inadequate surface runoff piping.

13. *GULLEYS, INCLUDING ALL SURFACE WATER ENTRANCES AND UPSTREAM ENDS OR HEAD WHERE MAINSTREAM ENTERS:* Check current rate of erosion; resulting pollution and sedimentation of downstream lakes, channels; damaged lands; impairment of bridges and other structures; need of erosion control such as temporary brush and wire dams and plantings.
14. *SPRINKLER SYSTEM NOZZLES, SPRAYS, HOSE, PIPE, AND VALVES:* Check for rust, corrosion, clogging, inadequate width or pressure, leakage, defective operation, evidence of water usage waste indicated by metering records or computations from nozzle-hours per acre per annum.
15. *FLOOD, IRRIGATION SYSTEMS, INCLUDING DELIVERY CHANNELS, GATES, FLOW-CONTROL AND WATER TURNOUT WORKS AND BORDER DIKES:* Check for defective operation, erosion, silting, scour, water loss, improper application, failure to supply to all parts of tracts.
16. *WINDBREAKS OF TREES:* Check for breakage, lack of vigor, dead or dying trees requiring replacement; disease and insect damage indicated from condition of leaves; branches interfering with utility lines; contour ridges in and sections inadequate to prevent surface runoff and retain and cause absorption of storm waters around tree.
17. *WEED CONTROL:* Vigor and rapid growth, indicating need for reapplying soil sterilents; erosion damage where soil sterilents were used; check for emergence of any and all

E	U	R	D

types of vegetation as an index of the efficacy of the remaining soil chemicals and of a need for applying additional chemicals to the soil; where selective contact sprays are used, check for percentage of kill and injury to vegetation that is to be preserved; check vegetation on adjoining lands for damage by spray drift.

Total Items: 17 Raw Total %

Findings (comment on each checkpoint and summarize to justify scoring):

___

___

___

Section 4. Railroad Trackage

SCOPE: Railroad trackage, including running or access tracks, classification yards, sidings, and storage tanks.

1. *TRACKS:*
   a. Vertical and horizontal misalignment caused by heaving, sinking, churning, inadequate expansion, particularly during hot weather: examine closely where track passes from earth fill to bridges or trestles. (Standard gauge is 4 feet, 8½ inches, which is increased on sharp curves.)
   b. Rough spots (ask train crews).
2. *RAILS:*
   a. Check for breaks, splits; cracks in head, web, or base; damage from flat wheels.

E	U	R	D

b. Check for creeping or shoving, particularly at curves or ends; battering, overflow, chipping.

c. *Joints:* check for loose angle or splice bars, loose and missing bolts, inadequate expansion. End of 33-foot rails should butt at 100°F. At lower temperatures, clearance of 1/32 inch for every degree difference or part thereof.

d. Loose spikes or failure to provide four per proper tie; improper tie plate seating (where used); improper support of rail.

e. Obstructed flangeways of girder-type rails; obstructed flangeways and tops not flush with pavements or crossing for standard-type rails.

f. *Road crossings:* poor condition, roughness to road traffic, obstructions.

3. *TIES:* Check for decay, splitting, general deterioration, rail cutting, insufficient or improper embedment in ballast to prevent movement, inadequate drainage.
4. *BALLAST:* Check for dirt and mud accumulations, soft or wet spots, grass or weeds, washing away and settlement, inadequate extension beyond ties; slope to grade steeper than 1 1/2 to 1.
5. *DRAINAGE:* Check for obstructed drainage ditches and culverts; erosion of side slopes and shoulders; actual or potential slides onto or close to track; washing or erosion at headwalls, inlets, discharge openings.
6. *TURNOUTS:* Check for lack of lubrication, clogged with debris or dirt, inadequately spiked; out-of-gauge; improper operating condition of switches, switch latches, targets, and lamps. (Correct throw at switch point is $4\frac{3}{4}$ inches.) Gauge from point of frog to flange face of guardrail where curvature through turnout

E	U	R	D

is 8° or less is 4 feet, 6⅝ inches, and in excess of 8° is 4 feet, 6¾ inches. Gauge on tangents and curves of 8° or less is 4 feet, 8½ inches, and for each 2° above 8° on curves, 1/8 inch is added to maximum of 4 feet, 9 inches.

7. *TANK CAR UNLOADING TRACKS: Bonding wires across rail joints, between rails and unloading header pipelines, and connections between rails and ground rods, and insulated rail joints:* loose connections, corrosion, frayed or broken strands. (Repair when 1 strand broken in 7-strand or more than 3 broken in 19-strand.)
8. *PROTECTIVE DEVICES:*
   a. *Warning Signs:* inadequate, improperly placed, illegible; telltales improperly placed and in poor condition; inadequate and poor structural stability of bumper blocks and cattle guards.
   b. *Guard Rails:* poor structural condition and improper placement at sharp curves, street embankments, trestles, bridges, or other locations where derailment would be serious.
   c. *Retaining Walls:* check for undermining, misalignment, weephole obstructions, or other deficiencies that would tend to endanger tracks.
9. *CLEARANCES:*
   a. Check if clearances at warehouses and structures less than 8 feet from center line of track from rail to 22 feet above rail; overhanging shed roof less than 5 feet, 6 inches from centerline of track from rail to 15 feet, 16 inches above rail; do not conform to AREA standards and are less than those required by servicing railroad.
   b. Check for presence of weeds or other obstructions blocking view, creating fire hazard, or reducing clearances.

E	U	R	D

Total Items: 9 Raw Total %

Findings (comment on each checkpoint and summarize to justify scoring):

____

____

____

____

____

____

____

____

____

____

Section 5. Pavements

SCOPE: Airfields, roads, streets, walks, parking areas for motor vehicles, and storage areas. Included are all types of pavements: rigid (concrete), asphalt-surfaced flexible pavements, and hard stands of stabilized earth materials of relatively low cost.

1. *ALL PAVEMENTS:*
   a. *Curbs, gutters, manholes:* Check for cracks, breaks, misalignment, damaged tops, inadequate expansion and crown.

E	U	R	D

b. *Drainage:* obstructed ditches, improper grading and shoulder protection.

2. *CONCRETE:*

a. *Expansion Joints:* not vertical; slot above expansion joint filler, not directly above filler; filler missing, unbonded, or extruded; dirt, sand, stone, or foreign material wedged in joint.

b. *Joints:* pumping evidenced by soil particles and water forced upward through joint; spalling caused by inadequate expansion joints or faulty construction or both.

c. *Cracks:* transverse cracks near doweled joints tend to be caused by poor dowel alignment. Interior corner cracking is indicative of poor subbase support or overloading. Longitudinal cracks are caused by improper joint spacing, inferior coarse aggregate in the concrete mix, or poor subgrade support and overloading. Radial cracks from a point are caused by overloading and/or "mushroom" earth support. Prismatic section of pavements, bounded by cracks, are also developed by overloading and irregular or nonuniform support. Cracks that crisscross and are extensions are caused by extensive overloading. "Cracked-out" sections that are pushed downward are usually caused by soft, wet subgrades covered with inadequate subbase. Such cracked-out sections may also be caused by nonuniform subsidence of the paved area. Corner or outer-edge cracking is caused by passage of moving loads from pavement to shoulder and back again.

d. *Depressions:* caused by consolidation of soft soil under load. May be evidenced by ponding of water after rains.

e. *Scaling:* usually caused by overfinishing, in which procedures provide an excess of ce-

E	U	R	D

ment in the pavement surface; sometimes caused by chemicals and ice used to remove snow and ice.

f. *Abrasion:* caused by blading equipment to remove snow and ice or to grade and reshape shoulders.

g. *Buckling:* caused when the pavement is restrained from expanding.

h. *Frost heave:* caused by expansion of a wet subgrade when it freezes.

3. *FLEXIBLE TYPE:*

a. *Grooving and shoving:* caused by high-pressure tires operating on surfaces of inadequate stability.

b. *Raveling:* aggregate particles pulling loose from surfacing.

c. *Burned areas:* heat effect from blasts of jet aircraft is cumulative. Surfaces become lifeless and brittle, separated aggregate may be blown about by blast.

d. *Softening:* caused by spillage of petroleum distillates or jet plane blast.

e. *Frost heave:* caused by expansion of a wet subgrade when it freezes.

g. *Depression:* caused by consolidation of soft soil under load. May be evidenced by ponding of water after rains.

h. *Weathering and oxidation:* asphalt becomes brittle, brown and dead in appearance. Irregular pattern of fine cracks in areas of little traffic appears in cold weather. Seal coating or rolling with rubber-tired roller required for surfaces that receive little or no traffic.

4. *BRICK AND STONE: High and low areas:* failure of bases, bedding courses, grouting; or loose or missing individual pieces.

5. *GRAVEL, CINDER, SHELL, AND STABILIZED SOIL:* breaks, potholes, corrugated,

E	U	R	D

rutting, inadequate crown, general deterioration.

Total Items: 5 Raw Total %

Findings (comment on each checkpoint and summarize to justify scoring):

____

____

____

____

____

____

____

____

____

____

Section 6. Retaining Walls

SCOPE: Retaining walls of all kinds, including cribbing and sheet piling when used as retaining walls. Waterfront structures are covered by other Inspection Guides.

1. *CONCRETE FOUNDATIONS:* Check for cracked, broken, scoured, spalling, exposed reinforcing; evidence of movement, settlement, and undermining.

E	U	R	D

2. *Concrete or Masonry Walls:* Check for cracked, broken, spalling, misplaced sections, general deterioration, exposed reinforcing, eroded and sandy mortar joints, bulging, vertical and horizontal misalignment.
3. *TIMBER WALLS AND CRIBBING:* Check for cracked, broken, loose, missing, wearing, undermining, rotting, insect infestation, bulging, vertical and horizontal misalignment.
4. *SHEET PILING AND BULKHEADS:* Check for rust, corrosion, bulging, vertical and horizontal misalignment.
5. Evidence of seepage resulting from obstructions in weepholes or other drainage outlets.
6. Loose or missing premolded expansion joint material allowing washout of backfill.
7. Structural inadequacy and poor physical condition
8. *Embankment slopes and areas behind walls:* erosion, settlement, or slippage resulting from improper drainage, lack of full sod or vegetation coverage, damage from burrowing animals, slopes steeper than angle of repose.

Total Items: 8 Raw Total %

Findings (comment on each checkpoint and summarize to justify scoring):

E	U	R	D

Section 7. Storm Drainage

SCOPE: Catch basins, curb inlets, pipelines, headwalls, outfalls, tide gates, drop structures and spillways, manholes, culverts, subsurface drainage, gutters, and ditches. A plan of the entire storm drainage system from inlets to outfall ditches should be available. Pumps and pump stations are covered by other checkpoints.

1. Ascertain that invert elevation on non-sedimentation basin pipe is the same.
2. *CATCH BASINS AND CURB INLETS:* Check for debris, obstructions, and cracked, broken, or improperly seated grating, settlement.
3. *PIPELINES:* Check for misalignment, settlement, cracked, broken, open joints, sediment, debris, tree roots, erosion in concrete pipes, erosion and corrosion in corrugated metal pipes. Inspect pipe smaller than 48 inches in

E	U	R	D

diameter by using a light between manholes. Crawl through pipe 48 inches in diameter or larger. Tightness of joints may be checked by blocking off a section between manholes for 24 hours to determine amount of ground water infiltration (watchman necessary to open line in event of rain).

4. *HEADWALLS:* Check for cracked, broken, spalling, exposed reinforcement, settlement, undermining condition of pipe joint at headwall.
5. *APPROACH CHANNELS:* Check for evidence of water channeling under and around pipe or headwall.
6. *OUTFALL AND CHANNEL BEYOND HEADWALL:* Check for sediment, debris, other obstructions, evidence of erosion of adjoining property.
7. *TIDE GATES:* Check for restricted or tight motion; loose closure; outfall line and bar screens for sediment and obstructions.
8. *DEEP STRUCTURES AND SPILLWAYS:* Check for silt accumulations and erosion.
9. *MANHOLE FRAMES AND COVERS:* Check for rust, corrosion, poor fit; ladder rungs for rust, corrosion, broken parts, damaged supports.
10. *MANHOLE WALLS:* Check for cracking, spalling, exposed reinforcement; eroded or sandy mortar joints, loose, broken, or displaced brick.
11. *MANHOLE BOTTOMS:* Check for clogging, restricted flow, silt, sewer pipe fragments (indicating broken pipe), invert elevation of outlet pipe not flush with bottom.
12. *CULVERTS:* Check for sediment, obstructions at inlets and outlets, ditch bottoms not flush with pipe inverts, and noncompacted or pervious soil resulting in channeling. (Cracks

E	U	R	D

in pavement over subsurface drainage and culverts indicate washout of soil from cracked or broken pipe, and obstructions are indicated by restricted flow after prolonged rainfall.)

13. *GUTTERS AND DITCHES:* Check for cracked, broken, eroded concrete surfaces, defective expansion joint, misalignment, obstructions, ponding of water, silting or sloughing-off of sides, inadequate side vegetation coverage to prevent erosion.
14. Standing water which would permit mosquito breeding in drainage system.

Total Items: 14 Raw Total %

Findings (comment on each checkpoint and summarize to justify scoring):

____

____

____

____

____

____

____

____

____

____

E	U	R	D

___

___

___

Section 8. Tunnels and Underground Structures

SCOPE: Tunnels of all kinds, including storage tunnels, pipeline tunnels, vehicular tunnels, and water tunnels; also, underground structures housing utilities, service installations, and similar equipment or operations. Not included are underground tanks or earth-covered storage bunkers that are wholly or partly above ground

1. *PORTAL STRUCTURES:* Check for drainage defects, cracks, breaks, leaks in face and between face and tunnel lining, eroded slopes or undermining.
2. *WING AND FACE WALLS:* Check for inadequate protection to personnel, erosion of slopes, loose rocks, actual or potential slides, scouting or undermining of walls.
3. *DOOR AND GATE OPERATING AND LOCKING DEVICES:* Check for rust, corrosion, loose, missing, or damaged parts, improper operation.
4. *CONCRETE FLOORS:* Check for cracks, breaks, scaling, other damage, surface dusting.
5. *EARTH AND GRAVEL FLOORS:* Check for improper grading and drainage, soft and muddy areas.
6. *VEHICULAR TUNNEL FLOORS:* Check for faulting at joints, scaling, abrasion, depressions, buckling.

E	U	R	D

7. Check tracks for misalignment, rails for damage and inadequate or loose connections and supports, and ties for rot or other damage.
8. *ALL LININGS:* Check for leakage, settlement, displacement. Concrete: cracks, breaks. Metal: looseness, rust, corrosion. Timber: cracks, breaks, rot.
9. *UNLINED TUNNELS:* Check for spalling, disintegration, loose or fallen rocks.
10. *METAL ROOFS:* Check for rust, corrosion, inadequate supports.
11. *PIPELINE TUNNELS:* Check for rust, corrosion, misalignment, broken, leakage.
12. Check for defective drainage system or facilities, particularly in storage tunnels.
13. *GENERAL CONDITION OR VENTILATION EQUIPMENT:* Check for apparent defects in operation rust, corrosion, loose, missing, or other damage to related parts.
14. *LIGHTING SYSTEMS AND FIXTURES:* Check for poor operating condition, inadequate, improper types.
15. *GROUNDING CONNECTIONS:* Check for electrical discontinuity, loose, missing, corrosion, or other damage to the connections. (Grounding of metal parts is required in tunnels where explosions may occur.)

Total Items: 15 Raw Total %

Findings (comment on each checkpoint and summarize to justify scoring):

____

____

____

E	U	R	D

Section 9. Piping Systems

SCOPE: Piping systems, both exposed and underground.

1. Inspect all exposed piping for leakage, corrosion, loose connections, and damage.
2. Inspect all underground piping for leakage, ponding, erosion, and settlement.
3. Check buried valves for bent stem, leakage corrosion, and proper operation.
4. Check exposed valves for bent stem, leakage, corrosion, and proper operation.
5. Inspect meters for accuracy, leakage, corrosion, and proper operation.
6. Inspect hydrants and hydrant shut-off valves (SOV) for missing caps, broken or missing chains, damaged threads, missing or damaged guards and identification markings.
7. Check hydrants for rust and corrosion.
8. Check valve and meter pit manholes and roadway boxes for rust, corrosion, and damage.
9. Check manhole frames, covers, and ladder rungs for rust, corrosion, loose or missing rungs, and other damage.

E	U	R	D

10. Check concrete and mortar joints in manholes.
11. Check walkways, guardrails, stairs, and ladders for rust, corrosion, broken or missing parts.
12. Check overflow pipes for water entrance, rust, and damage to screen.

Total Items: 12 Raw Total %

Findings (comment on each checkpoint and summarize to justify scoring):

____

____

____

____

____

____

____

____

Section 10. Steam Distribution

SCOPE: Cover equipment included in steam distribution lines.

1. Check operation and condition of pressure safety relief valve.
2. Check Y-strainer.
3. Check basket strainer.

E	U	R	D

4. Inspect 8-inch single or double (traverse) sliding expansion joint (injection packed).
5. Inspect thread gland expansion joint (up to 3-inch diameter).
6. Inspect and repack single or double (traverse) flanged expansion joint.
7. Check air or water separator, float type with 1-inch I.D. fittings.
8. Check thermostatic steam trap.
9. Check drip pocket (2 inches). Inspect bucket steam trap.
10. Inspect bucket team trap (high volume, upright float).
11. Inspect float and thermostatic steam trap (up to 2-inch diameter).
12. Inspect vacuum valve, or balancing arm pressure reducing valve.
13. Inspect check valve (2 to 4 inches).
14. Inspect pressure-regulating valve.
15. Check for water in mechanical room.
16. Check unions for leaks or cracks.
17. Steam traps:
    a. Check traps and bypass valves for leakage, damage, and proper operation.
    b. Check strainers for leakage, damage, proper operation, and clogged screens.

Total Items: 17 Raw Total %

Findings (comment on each checkpoint and summarize to justify scoring):

____

____

____

____

E	U	R	D

Section 11. Sump Pump

SCOPE: Covers sump pump, condensate or vacuum pumps, vacuum producer pumps, and liquid transfer pumps.

1. *SUMP PUMP:*
   - *a.* Check pit for water level.
   - *b.* Check bail, floats, rods, and switches. (Make sure float operates as designed.)
   - *c.* Inspect motor and pump.
   - *d.* Inspect check valves.
2. *CONDENSATE OR VACUUM PUMP:*
   - *a.* Operate unit to check for steam binding.
   - *b.* Check condensate temperature (should be approximately 30° below steam temperature if traps are not leaking).
   - *c.* Examine flanges for steam leaks.
   - *d.* Check receiver.
   - *e.* Check motor float switch and float operation on high, low water level. Inspect pressure switches.
   - *f.* Examine receiver, vent pipe, inlet and discharge openings for excessive corrosion.

E	U	R	D

- *g.* Check alignment of coupling with straight edge.
- *h.* Check lubrication of pump and motor.
- *i.* Examine vacuum breaker operation.
- *j.* Inspect ball floats, rods and other linkage.

3. *VACUUM PRODUCER PUMP:*
   - *a.* Inspect wiring and electrical controls for loose connections, charred, broken or wet insulation, short circuits, etc.
   - *b.* Check lubrication of motor and/or pump as applicable.
   - *c.* Check motor for excessive heat and vibration.
   - *d.* Inspect for rust and corrosion.
   - *e.* Check alignment and condition of belts.

4. *LIQUID TRANSFER PUMPS:*
   - *a.* Check with operating personnel for any known deficiencies.
   - *b.* Check pump and motor exterior for cleanliness.
   - *c.* Check for corrosion on pump exterior and base plate.
   - *d.* Check for leaks on suction and discharge piping, seals, packing glands, valves, etc.
   - *e.* Check pump operation: vibration, noise, overheating, etc.
   - *f.* Check alignment, clearances and rotation of shaft and coupler (includes remove and install safety cover).
   - *g.* Check for loose, missing, or damaged nuts, bolts, or screws.
   - *h.* Check pump and motor for lubrication.
   - *i.* Check electrical system (motor and motor starter) for loose connections and frayed wiring.
   - *j.* Check suction or discharge pressure gauge readings and flow rate.
   - *k.* Check packing glands. Note that slight dripping is required for proper lubrication of shaft.

E	U	R	D

l. Inspect float assembly for proper operation and clean sump pump (sump pumps only).

Total Items: 4 Raw Total %

Findings (comment on each checkpoint and summarize to justify scoring):

____

____

____

____

____

____

____

____

____

____

____

Section 12. Fresh Water Supply and Distribution System

SCOPE: Underground and surface water supplies, intakes, piping valves, valve pits, roadway boxes and manholes, and meters. Pumps are covered by other checkpoints.

E	U	R	D

1. *EXPOSED PIPING:* Check for leakage, corrosion, loose connections, defective caulked joints on bell and spigot pipe, loose bolts on flanged pipe and clamp-type connections, damaged or missing hangers and supports, mechanical damage, damage to protective coating.
2. *UNDERGROUND PIPING:* Check for evidence of leakage, ponding, erosion, or settlement or areas adjacent to piping; excessive supply pressure, water hammer or vibratory noises in line. (When exposed for alteration or repair, examine for deficiencies similar to those listed for Exposed Piping.) (Breaks may be found by oral leak indicator.)
3. *BURIED GATE VALVES:* Check for damaged operating nuts, bent stem, valve opens clockwise, difficult to operate when partially closed and then opened wide, location tie-in error.
4. *UNBURIED GATE VALVES:* Check for leaks, corrosion, visible defects in stem, operating handwheel, lever, body, packing gland, flanges, and gaskets, damage to protective coatings, difficult to operate when partially closed and then opened wide, location tie-in error.
5. *CHECK VALVES:* Check for leakage, rust and corrosion, other damage.
6. *VACUUM AIR RELIEF VALVES OR BYPASS VALVES:* Check for leakage, corrosion, defective operation, unauthorized or doubtful cross-connections in bypass valves.
7. *BLOWOFF VALVES:* Check for leakage, corrosion, defective operation, plugged drain.
8. *DOUBLE-CHECK VALVES:* Check for corrosion, leakage, tightness, failure to prevent back flow.
9. *REDUCED PRESSURE VALVES:* Check for rust, corrosion, loose connections, water

E	U	R	D

draining continuously from relief valve opening, clogged drain.

10. *TEST REDUCED PRESSURE VALVES:* Check for pressure in zone between check valves—should be at least 2 psi less than supply pressure. Check valves must be tight against reverse flow under all pressure differentials.
11. *METERS:* Check for leaks, corrosion, broken glasses, moisture behind glasses, settlement, evidence of fault operation, records indicate periodic test not performed.
12. *HYDRANTS AND HYDRANT SHUT-OFF VALVES (SOV):* Check for missing caps, missing or broken chains, damaged or worn nozzle threads and operating nuts, missing protective guards or identification markings, corrosion, deteriorated paint, evidence of improper wrenches being used on operating nuts.
13. *TEST HYDRANTS AND SHUT-OFF VALVES:*
    - *a.* Do not flush in cold weather; starting with hydrant nearest source, slowly open hydrant to wide-open position and let flow until reasonably clear, then shut tight. If water comes up around hydrant while wide open, drain-valve facing or gasket must be replaced; examine barrel for cracks and leakage past caps; if top of hydrant leaks, remove clover, tighten packing gland or repack.
    - *b.* Close hydrant and shut-off valve; water spillage from nozzle indicates hydrant valve and seat or SOV leakage; noisy valve indicates leakage; quiet valve and stable water level indicate drain-valve stoppage.
    - *c.* Where groundwater table is above hydrant, drain and pump out barrel; lower

E	U	R	D

small weight on string for evidence of ice in barrel.

14. *VALVE AND METER PIT MANHOLES AND ROADWAY, BOXES:* Buried, protruding, over too close to stem, poor fit or missing covers, clogged vent holes, rust, corrosion, rot, cracked, split, and structural damage to individual members, improper grading of adjacent surrounding areas, resulting in diversion of surface water into enclosures, debris or other accumulations, errors in location tie-ins.
15. *MANHOLES FRAMES, COVERS, AND LADDER RUNGS:* Check for rust, corrosion, loose, broken, missing, rot, splintered, other damage to individual parts.
16. *MANHOLES:* Check for cracked, broken, spalling, deteriorated mortar joints, other damage, improper grading of adjacent surrounding areas, resulting in diversion of surface water into manholes, need to raise manhole cover to or above existing grade.
17. *HEADWALLS:* Check for leakage, cracks, spalling, displaced sections, settlement, defective mortar or expansion joints, soil erosion at footings, damaged screens and poured-in-place pipe connections.
18. *WALKWAYS, GUARDRAILS, STAIRS, AND LADDERS:* Check for rust, corrosion, loose, broken, rot, missing, settlement, wooden members in contact with ground surfaces.
19. *DAMS AND SPILLWAYS:* Check for leakage, cracks, spalling, settlement, erosion at abutments, excessive debris, lack of provision for tipping or carrying way of flashboards in event of topping.
20. *CONCRETE, MASONRY, AND STEEL CURBINGS:* Cracks, spalling, loose, displaced, or defective joints, rust, scale, other damage.

E	U	R	D

21. *AREAS SURROUNDING OR ADJACENT TO WELLS, SPRINGS, WATER STORAGE, BASINS, WATERSHEDS, CATCHMENTS, SHORELINES, AND EMBANKMENTS:* contamination or pollution, erosion, vegetation, improper soil cover, settlement, flooding, burrowing animals, decay, aquatic weed growth, algae, organic deposits, industrial wastes; sewage, gas, oil, and chemical spillage.
22. *REDUCTION IN WATER YIELD IN WELLS, INFILTRATION GALLERIES, AND ALL STORAGE, TREATMENT, AND COLLECTION BASINS:* check for opening in casings, imperviousness of soils and porous linings, leakage, clogged intake screens, continued dry spells, defective well pump operation.
23. *OVERFLOW PIPES:* Installed below surface water entrance, rust damage to screens.

Total Items: 23 Raw Total %

Findings (comment on each checkpoint and summarize to justify scoring):

____

____

____

____

____

Section 13. Sewage Collection and Disposal Systems

SCOPE: Installed sewage collection and disposal systems.

E	U	R	D

1. Inspect grease traps, oil interceptors, and similar equipment for accumulations of scum and grit, and operation.
2. Inspect frame, cover, and ladder rungs for rust, corrosion, fit of cover, missing, damage.
3. Inspect concrete and masonry for cracks, breaks, spalling, deteriorated mortar joints.
4. Inspect piping for corrosion, open joints, cracked or crushed sections, obstructions.
5. Inspect inverted siphons and depressed sewers for clogging, sluggish flow, accumulations of grit and debris.
6. Check condition for leakage, rust, corrosion, deteriorated coatings.
7. Inspect supports and anchors.
8. Check bar screen and raker for corrosion and damage.
9. Inspect cutters.
10. *CHLORINATOR:*
    a. Inspect housing for adequate ventilation.
    b. Inspect assembly for rust, corrosion, leaks.
11. *VEGETATION AND ADJACENT GROUNDS:*
    a. Inspect grass and ground cover plants.
    b. Inspect ground surfaces for indications of seepage from sewers.
12. *TIDE GATE:*
    a. Check operation.
    b. Inspect for blockage.
13. *SPECIAL INSPECTION DURING OR AFTER PROLONGED RAIN OR SEVERE STORM:*
    a. Inspect manholes for infiltration, proper grading.
    b. Inspect underground piping.
    c. Inspect aboveground piping.
    d. Inspect vegetation on slopes for stability.

Total Items: 13 Raw Total %

E	U	R	D

Findings (comment on each checkpoint and summarize to justify scoring):

____

____

____

____

____

____

____

____

____

Section 14. Unfired Pressure Vessels

SCOPE: Unfired pressure vessels except (a) cylinders for shipment of compressed or liquefied gasses; (b) air tanks for brakes or vehicles; (c) unfired pressure vessels having a volume of 5 cubic feet or less: (d) unfired pressure vessels designed for a working pressure not exceeding 15 pounds per square inch gauge; (e) unfired pressure vessels containing only water under pressure for domestic supply purposes, including those containing air, the compression of which serves only as a cushion; (f) unfired pressure vessels used as refrigerant receivers for refrigeration and air conditioning equipment; (g) expansion or accumulator tanks used in conjunction with high temperature water installations.

E	U	R	D

Previous Inspection Reports: indicate decrease in pressure-carrying capacity; recommendations in previous reports not completed or not scheduled for completion.

1. *EXTERNAL INSPECTION:*
   - *a. Safety and relief valves:* Check for accumulated rust, scale, or other debris; obstructed drain; hazardous conditions created by discharge; try lever not free.
   - *b. Rupture disks:* Check for broken, leaking, deteriorated; plugged vent.
   - *c. Pressure-indicating gauges:* Check for broken, missing, or dirty glass; illegible markings; bent pointer; leaking connections; inoperative.
   - *d. Lagging:* Check for loose or missing material; crack, open seams, evidence of vapor or water leaks.
   - *e. Shell:* Check for corrosion, leakage, cracks, distortion, cracked or broken welds; loose or broken rivets; loose or missing caulking.
   - *f. Supports:* Check for settlement, deterioration, lack of rigidity; cracks, loose or dislodged matter, excessive corrosion, cracked or broken welds, loose or missing bolts or rivets, warped or bent structural members.
   - *g. Piping:* Check for leakage, strain or torsion, excessive corrosion, improper drainage, misalignment, lack of support, inadequate provision for expansion or contraction, excessive vibration, pockets at valves and connections, settlement,
   - *h. Stop and check valves:* Check for loose, missing, broken parts; excessive wear or corrosion, leakage, obstructed drain openings.
   - *i. Pressure control switch:* Check for loose, missing, broken parts or connections, corrosion, rust or other substance preventing proper operation.

E	U	R	D

2. *INTERNAL INSPECTION:*
   a. *Preparation for inspection:* Inadequate, incomplete, untimely.
   b. *Pressure gauge:* Improperly calibrated.
   c. *Vessels secured or stored:* Not completely dry; inadequate supply of desiccant; improper or inadequate placement of desiccant.
   d. *Corrosion-resistant lining:* Check for cracks, corrosion behind lining, corrosion behind deposits on lining.
   e. *Minimum thickness (surfaces exposed to corrosion from vessels' contents):* Determine.
   f. *Shell Plates:* Check for cracks, defective joints, distortion, erosion, corrosion; grooving, lap seam cracks; cracked or severely corroded rivets; cracked welds. Determine minimum thickness and calculate new maximum allowable working pressure.
   g. *Heads:* Check for cracks, deformation, excessive corrosion.
   h. *Stays and braces:* Check for cracks, bends, looseness, or erosion; excessive tension, excessive corrosion or erosions; loose, cracked, broken connections.
   i. *Nozzles:* Check for distortion, excessive corrosion; cracked welds; loose, cracked, or severely corroded rivets or bolts; poor or ineffective threaded conditions
   j. *Reinforcing plates:* Cracks, deformations, other weaknesses.
   k. Openings and connections for piping and external attachments: Obstructed; inadequate; excessive corrosion.
3. *HYDROSTATIC OR PNEUMATIC TEST:*
   a. *Preparation:* Inadequate, incomplete.
   b. *Inspection:* Significant drop in pressure in 16 minutes; leakage, loose parts, deformation.

4. *INSPECTION OF OPERATION:*
   a. *Controls:* Inability to maintain proper pressures; improper adjustment for cutout and cut-in devices.
   b. *Piping:* Check for leakage, excessive vibration, tendency to crystallize.
   c. *Pressure-indicating gauges:* Stuck pointer, restricted movement of pointer, obstructed connections.
   d. *Temperature-indicating devices:* Check for temperatures indicated, particularly immediately after high load demands.
   e. *Stop and check valves:* Excessive vibration, ineffective or defective operation.
   f. *Pressure-reducing valves:* Defective, inadequate, improper operation.
   g. *Metering and recording devices:* Improper operation.
   h. *Safety and relief valves:* Improper operation, obstructed discharge, do not release at required pressure.

E	U	R	D

Total Items: 4 Raw Total %

Findings (comment on each checkpoint and summarize to justify scoring):

____

____

____

____

____

____

E	U	R	D

Section 15. Underground Tanks

SCOPE: Subsurface tanks, tank enclosures, and tank fittings and appurtenances.

1. *FOUNDATIONS:* Check for settling, movement, upheaving, inadequate soil coverage.
2. *STRUCTURAL SUPPORTS AND CONNECTIONS:* Check for rust, corrosion, rot, broken, cracked, distorted, loose, missing, deteriorated paint.
3. *TANK LINING:* Check for loss of elasticity, granulation discoloration, cracks, peeling, sloughing off.
4. *TANK INTERIOR:* Check for rust, corrosion, ale, teriorated protective coatings.
5. *FRAMES AND COVERS ON MANHOLES AND HATCHES:* Check for rust, corrosion, cracks. breaks, missing or damaged bolts, worn or defective hinges and gaskets.
6. *VENTS:* Check for rust, corrosion, dirty screens.
7. *PRESSURE AND VACUUM RELIEF VALVES:* Check for defective operation, leakage, improper adjustment.
8. *MANOMETERS AND THERMOMETERS:* Check for inaccuracy, mechanical damage, loss of fluid.
9. *FLOAT GAUGES:* Check for wear, binding, apparent inaccuracy.
10. *ROOF DRAINS AND SCREENS:* Check for missing, rust, clogging.
11. *GROUND CONNECTIONS:* Check for loose, missing mechanical damage; corrosion interfering with electrical continuity.

12. *INTERIOR HEATING, INLET AND OUTLET PIPES, NOZZLES, SUPPORTS, SUMPS AND SUMP DRAINS:* Check for rust, corrosion, wear, loose or missing parts.
13. *DIKES:* Check for cracks, breaks, spalling, rust, corrosion, settlement, heaving, soil erosion, water seepage; inadequate sod cover on outer face where earth-filled; inadequate treatment of inner face to prevent vegetation growth; access steps for settlement, breaks, other damage.
14. *DRAINAGE DITCHES SUMPS AND EARTH SURFACES BETWEEN DITCH AND FOUNDATION:* Check for improper slope to divert surface water away from foundation and berm; trash and debris; erosion.
15. *LEAKAGE:* Review of records indicates fuel losses; water in oil samples.
16. Locate leaks by filling completely with water and applying hydrostatic pressure of 4 feet of water for not less than 4 hours.

E	U	R	D

Total Items: 16 Raw Total %

Findings (comment on each checkpoint and summarize to justify scoring):

____

____

____

____

____

____

E	U	R	D

Division

IV

Utilities Plants

E	U	R	D

I. *OBJECTIVE:* The basic objective is to maintain utilities plants in a functional and economical manner that will be consistent with operating requirements and sound engineering practice. Also ensure reliability and continuance of service consistent with the authorized mission of the organization.

II. *DEFINITIONS:*

A. Utilities Plants are classified as follows:

1. Fresh water supply.
2. Electrical generation and/or distribution.
3. Heating generation.
4. Air-conditioning generation and distribution

B. Sections: Components for which assessment checklists are established are listed in II.A 1-4, above.

III. *MAINTENANCE STANDARDS:* All utility plants shall be maintained to the extent necessary to ensure reliability of service, the full safety of operators, efficient operation of the plant, and prevention of unwarranted deterioration of equipment.

Section 1. Fresh Water Supply

E	U	R	D

SCOPE: Fresh water is normally purchased from a local utility company. These checkpoints cover a fresh water storage facility and water meters, as applicable.

1. *FRESH WATER STORAGE:*
   a. Check foundations for settlement, damage, rot, insect infestation.
   b. Check steel tanks for rust, corrosion, leakage, scale, damage protective coating, and damage.
   c. Check concrete tanks for damage, cracks, spalling, leakage, and damaged protective coating.
   d. Check wood tanks for leakage, rot, insect infestation, and damage.
   e. Check towers for rust, corrosion, damage, rot, and insect infestation, as applicable.
   f. Check expansion joints for loose or missing sealant.
   g. Check earth embankments for erosion, ponding of water, and leakage.
   h. Check valves, piping, fittings, and sleeves for rust, corrosion, leakage, and damage.
2. *WATER METERS:*
   a. Check meter installations for rust and erosion.
   b. Check meter for leaks.
   c. Check for refuse and dirt in the meter pit.
   d. Test meter for proper operation.

Total Items: 2 Raw Total %

Findings (comment on each checkpoint and summarize to justify scoring):

____

____

E	U	R	D

Section 2. Electrical Generation and/or Distribution

SCOPE: Electricity is normally purchased from a local utility company; however, checkpoints are provided for a large electrical power plant. Also provided are checkpoints for power distribution equipment and emergency electrical generators.

1. *LARGE ELECTRICAL POWER PLANT, GENERAL:*
   a. Check for housekeeping; lack of cleanliness or orderliness.
   b. Safety signs and posted instructions: inadequate, illegible, improper location.
   c. *Operating log, plant log, and maintenance records:* Failure to record pertinent readings and other information necessary to locate and evaluate trouble areas and trends.
   d. Evidence of need to follow up deficiencies that may lead to breakdown.
2. *GENERATOR LOADING:* Operating log: review or duration and amount of overload, am-

E	U	R	D

bient temperature, temperature rise. Note when rated temperature is approached or exceeded.

3. *GENERAL INSULATION:* Make insulation test measurements of generator field, armature windings, and cable from main breaker terminals. Note evidence of electrical weakness to extent that normal operating voltage or surges may result in failure.
4. GENERATOR EXCITATION SYSTEMS:
   a. Inadequate, not serviceable, unreliable.
   b. Poor physical condition of emergency exciters and associated equipment, including rheostats, pilot exciters, voltage regulators, motor devices.
   c. Inadequate ground indicating system in ungrounded exciter circuits.
5. *PLANT BATTERY:*
   a. Battery room or enclosure: Lack of cleanliness, unacceptable temperature, inadequate ventilation, unsatisfactory condition of floor, fire hazard from, lighting and power fixtures, fittings, and cable.
   b. Operating and maintenance records: Review for deficiencies in specific gravity levels, cell temperature, makeup water history, equipment.
   c. Connections: Loose, corroded, dirty, inadequate.
   d. Cells: Oversulphated plates, physical erosion, internal shorts, buckled plates, cracked grids, dirty electrolyte, improper electrolyte, level, excessive sedimentation.
   e. Chargers and controls: Poor physical condition.
   f. Instruments: Inaccurate (check frequency of calibration).
6. *BUSES:* Poor condition, dirty, structural distortion, loose joints and connections, evidence of overheating.

	E	U	R	D
7. *CONTROL SWITCHBOARDS:*				
a. Poor physical condition, dirty.				
b. Wiring and connections: Lack of neatness, looseness, corrosion.				
c. Fuses in control wiring system: Improper size.				
d. Indicating lamps: Not operating.				
e. Mimic buses: illegible, inaccurate.				
8. *EXTERIOR POWER DISTRIBUTION EQUIPMENT:*				
a. Inspect small dry core transformer less than 600V, including brushing exterior, check connections (two men).				
b. Inspect oil-type transformers, checking and recording oil levels and temperatures, and taking an oil sample (two men).				
d. Inspect oil-type transformer, pole mounted, checking and recording oil levels and temperature and taking an oil sample (two men).				
e. Gain access to substation.				
f. Open and close three phase switchgear circuit breaker including calling distribution dispatcher and rolling breaker out and in.				
g. Install and remove grounding cluster from ground using shotgun stick.				
h. Check bushings on high-voltage equipment. CAUTION: Make certain that equipment is deenergized and, if possible, grounded.				
i. Open, or close, air gap gang switch.				
j. Open, and close, access doors or panels to high-voltage buss bars.				
k. Check high-voltage equipment for charge using a statiscope after equipment is disconnected.				
l. Crank out (down) and crank in (up) voltage circuit breaker-distribution switch.				
9. *EMERGENCY GENERATORS UP TO 15 KVA:*				

E	U	R	D

a. Check oil level.
b. Check battery water level. Check terminals for corrosion.
c. Test run generator. Check for proper operation and switching while running.
d. Check belts for proper tension and wear.
e. Check fuel level in tank with gauge pole.
f. Check air filter.
g. Check engine oil.
h. Check spark plug condition.
i. Check wiring, connections, switches, etc.

10. *EMERGENCY GENERATORS GREATER THAN 15 KVA:*
    a. Check oil level in crankcase with dipstick.
    b. Check radiator water level.
    c. Check battery water level. Check terminals for corrosion.
    d. Test run generator. Check for proper operation.
    e. Check belts for wear and tension.
    f. Check fuel level in underground tank with gauge pole.
    g. Check air filter.
    h. Check diesel engine oil.
    i. Check gasoline engine oil.
    j. Check oil filter.
    k. Check spark plug condition.
    l. Check electrical wiring, connections, switches, etc.

Total Items: 10 Raw Total %

Findings: (comment on each checkpoint and summarize to justify scoring):

____

____

E	U	R	D

____

____

____

____

____

Section 3. Heating Generation

SCOPE: Boilers, expansion drums on high-temperature water installations, boiler auxiliaries and controls, and other heating generation equipment, i.e., heat pumps; heater equipment, forced air, oil, or gas; heater and controls console; and heating and ventilating unit.

1. *BOILERS:* Previous Inspection Reports: indicate decrease in pressure-carrying capacity; recommendations in previous reports not completed or not scheduled for completion.
2. *EXTERNAL INSPECTION:*
   a. *Safety and relief valves:* Check for accumulated rust, scale, or other debris; obstructed drain; hazardous conditions created by discharge; try lever not free.
   b. *Automatic low water (level or flow) fuel cut-off and / or water feeding device:* Check for rust, corrosion, deteriorated or defective parts, improper functions.
   c. *Gauges:* Check for cracked, broken, missing or dirty glass, illegible markings; bent pointer; leaking connections; improper function of cock between gauge and boiler.
   d. *Water columns and gauge glasses:* Excessive corrosion, cracked or dirty glasses, leakage, improper drainage.

E	U	R	D

e. Material storage: Lumber or material boiler or setting.

f. Casing: Distortion, slippage of bricks, open seams, cracks, looseness.

g. Shell: Check for corrosion; cracks, leaking roofs, valves, pipes, rust streaks on covering.

h. Boiler doors: Check for sagging, warping, cracking, chipped or broken edges, worn hinges, defective locks or latches; improper operation, deterioration, or damaged blast deflectors.

i. Breechings: Check for excessive corrosion, cracked welds, loose or broken connections, separated sections.

j. Overhead machinery: Check for loose parts or material that may drop on or strike boiler.

k. Foundation: Check for settlement, improper level.

l. Piping: Check for leakage, strain or torsion, excessive corrosion, improper drainage, misalignment, lack of support, inadequate provision for expansion or connections, settlement, improper tension, and alignment in supports.

m. Stop and check valves: Check for loose, missing, broken parts, excessive wear or corrosion, leakage, obstructed drain openings.

n. Pressure-reducing valves: Loose, missing, broken parts, rust, scale, other substance preventing proper operation.

o. Blow-off tanks: Excessive corrosion, cracks, distortion, other weakness; leaks, water pockets; improperly placed valves.

r. Ladders and runways: Broken, cracked, split, badly worn members; excessive corrosion; loose or missing bolts or other con-

E	U	R	D

nections; broken welds; abnormal deflection; loose or warped sections; slippery, surfaces; inadequate anchorage.

s. Electrical steam generators: Loose connections, burnt, corroded, frayed, or broken strands in grounding cable; loose connections, deteriorated screens or guards; missing, illegible, or improperly posted warning signs.

3. *INTERNAL INSPECTION:*

a. Preparation for inspection: Inadequate, incomplete, untimely.

b. Steam gauge: Improperly calibrated.

c. Boilers secured or stored: Wet layup: incompletely filled; improper pH and sulfite concentration. Dry layup: not completely dry; inadequate supply of desiccant; improper or inadequate placement of desiccant.

d. Waterside metal surfaces: Check for evidence of oil; indications of scale. Report presence of oil. Report location, amount, density, degree of hardness, and age of scale.

e. Plates: Check for cracks, defective joints distortion, erosion, excessive corrosion; grooving; lap seam cracks; cracked or severely corroded rivets; cracked welds. Determine minimum thickness and calculate new maximum allowable working pressure.

f. Stays and braces: Check for cracks, bends, looseness, uneven tension, excessive corrosion or erosion; loose, cracked, broken connections.

g. Reinforcing plates: Check for excessive corrosion, worn, cracked, deformed, insecurely fastened, openings obstructed.

h. Openings and connections for piping and external attachments: Obstructed; inadequate; excessive corrosion.

E	U	R	D

i. Internal piping and fittings: Check for loose connections, breaks, cracks, excessive corrosion, clogging.

j. Protection for blow-off connection: Check for spalling, cracking, looseness; missing or worn parts.

k. Manholes and handholes: Check for corrosion, leakage; distorted, elongated, excessively corroded holding clamps.

l. Tubes: Check for deterioration, excessive reduction in thickness, bulges, cracks, defective welds, erosion, corrosion, waste pockets, scale distortion.

m. Ligament: Check for broken, cracked, leaks.

n. Drumheads: Check for cracks, deformation, excessive corrosion.

o. Flaming: Grooving.

p. Soot blowers: Check for worn, loose, or inadequate bearings and parts; incorrect alignment of nozzles; excessive scouring of refractory baffles, incorrect blowing ranges; inadequate traps.

q. Fireside or gas-side metal surfaces: Check for steam pockets, corrosion, bulging, blistering, distortion, deformation, excessive burning action, fly ash deposits, flame leakage, localization of heat, deteriorated or loose tie rods and buck stays.

r. Fusible plugs: Check for poor condition; unsound metal; in service more than one year; exposed to steam temperature in excess of 425°F.

s. Setting: Check for improper level, tendency toward settlement; inadequate provision for expansion and contraction; excessive corrosion; holes; cracks; slippage; distortion; binding; loose, missing, deteriorated, improperly placed baffling.

E	U	R	D

t. *Furnaces and baffles:* Check for spalling, cracking, settlement, distortion, abnormal cracks and seams, excessive burning and corrosion, fly ash deposits, inadequate expansion joints, improper protection of mud drums; loose, missing, deteriorated, improperly placed baffling.

u. *Firing equipment and dampers:* Burners: loose, broken, or missing parts; excessive wear or corrosion. Grates, stokers and feeders: loose, broken or missing parts; excessive wear or corrosion. Dampers: loose, broken, missing, warped, or binding parts; excessive wear or corrosion; improper operating condition.

v. *Low-water fuel cut-off and / or water feeding device:* Check for scale or dirt deposits; restricted moving parts; excessive wear; loose, broken, or missing parts; improper function.

4. *HYDROSTATIC TEST:*
   a. *Preparation:* Inadequate, incomplete.
   b. *Inspection:* Significant drop in pressure after 15 minutes; check for leakage, loose parts, distortions, deformations. Report row and number of tubes requiring rolling or replacement.
   c. *Maximum allowable working pressure:* Recalculated.
5. *INSPECTION OF OPERATION:*
   a. *Firing equipment:* Improper or inefficient operation.
   b. *Controls:* Inability to maintain proper steam pressure or water temperature and air fuel ratio throughout capacity range of boiler with load swings of the rapidity encountered in operation; improper programming sequence and timing; improper cutoff of fuel supply; inability to maintain proper

E	U	R	D

water level or to operate cutoff devices. Test flue gas for CO^2 and O^2 levels.

c. *Steam gauges:* Stuck pointer; restricted movement of obstructed connections.

d. *Water columns and gauge glasses:* Restricted connections.

e. *Steam gauges:* Stuck pointer; Restricted movement of pointer; obstructed connections.

f. *Temperature-indicating devices:* Excessive temperatures indicated, particularly immediately after high load demands.

g. *Blow-off valves:* Restricted openings; excessively worn or otherwise defective.

h. *Stop and check valves:* Excessive vibration ineffective or defective operation.

i. *Pressure reducing valves:* Check for defective, inadequate, improper operation.

j. *Metering and recording devices:* Improper operation.

k. *Boiler auxiliaries:* Steam leakage, wastage to atmosphere, unnecessary use, inadequate or improper functioning.

l. *Boiler safety and water pressure relief valves:* Improper operation, obstructed discharge, does not release at required pressure. Adjust to proper setting requirements.

m. *Feedwater treatment:* Equipment: ineffective, inadequate, improper operation. Materials: incorrect, insufficient.

n. *Fuel-handling practices:* Multiple handling, use of duplex equipment for small loads, unnecessary heating, improperly maintained equipment, inefficient operation.

o. *Partial loading:* Unnecessary use of similar equipment at part load when one unit could carry load.

E	U	R	D

6. *HEAT PUMPS:*
   a. Inspect piping for evidence of leaks and vibration.
   b. Inspect all wiring for deterioration. Check for corrosion.
   c. Check mounting bolts.
   d. Check crankcase heater.
   e. Check fan for vibration or excessive noise.
   f. Check refrigerant levels. Check for leaks if loss of refrigerant is detected.
   g. Check temperature drop across condensing coil.
   h. Check air intake and screens for cleanliness.
   i. Check coil surfaces.
   j. Check that reversing valve is energized in the "heat" mode and deenergized in the "cool" mode.
   k. Check oil.
7. *HEATING EQUIPMENT, FORCED AIR, OIL, OR GAS:*
   a. Check combustion chamber for gas and oil leaks.
   b. Inspect electrodes and nozzles on oil burners. Inspect fire.
   c. Inspect fuel system for leaks.
   d. Check fuel strainer element on oil burner.
   e. Check fuel level in tank. Check tank, fill pipe, fuel lines and connections for leaks.
   f. Check for proper operation of pilot and burner primary controls. Check thermostat.
   g. Remove and reinstall filter panel or blower access panel.
   h. Inspect air filter in air handler.
   i. Check blower and motor for alignment, vibration, and lubrication.
   j. Check belts for wear and proper tension.
   k. Check burner operation through complete cycle or up to 10 minutes.

E	U	R	D

l. Check electrical wiring to burner controls and blower.
m. Check fire box for cleanliness.
n. Check blower and air plenum.
o. Check condition of flue pipe, damper, and stack.
p. Check building boiler gas pressure regulator vents for cleanliness and leaks. Check gas pressure regulator by using water column pressure gauge.

8. *HEATER AND CONTROLS CONSOLE:*
a. Check air line lubricator.
b. Check lines to and from unit for steam, water, or oil leaks.
c. Check dust filter for dust accumulation.
d. Check alignment of pulleys and tighten set screws.
e. Check condition of belt.
f. Check electric motor bearing for lubrication.
g. Check steam valves for proper operation.
h. Check steam trap.
i. Check motor bearings for noise or wear.
j. Check base bolts for tightness.
k. Inspect wiring and electrical controls for loose connections; charred, broken, or wet insulation; evidence of short circuiting and other deficiencies.

9. *HEATING AND VENTILATING UNIT:*
a. Check air filter
b. Check heater operation.
c. Check and inspect belt.
d. Check shaft and motor bearings or lubrication.
e. Check wiring and electrical controls for loose connections; charred, frayed, or broken insulation.
f. Check for rust and corrosion.

Total Items: 9 Raw Total %

E	U	R	D

Findings (comment on each checkpoint and summarize to justify scoring):

Section 4. Air Conditioning Generation and Distribution

SCOPE: Air conditioning, ventilating, and related equipment, including refrigeration units, evaporative coolers, and condensers.

1. *COMPRESSOR, 3–100 TON:*
   a. Check compressor operation. Observe refrigerant in sight glass while compressor is operating.
   b. Turn power off and on.
   c. Remove and install access panel.

E	U	R	D

d. Check piping and valves for refrigerant leaks.
e. On open compressors, check belt for wear, proper tension, and alignment.
f. On open compressors, check shaft bearings and motors for lubrication.
g. Check electrical controls for loose connections and frayed wires.
h. Check crankcase oil temperature and pressure gauges. Record readings.
i. Check for proper refrigerant charge and readings.
j. Pump compressor down into receiver.
k. Charge compressor.
l. Check unit for corrosion.
m. Check unloaders or proper operation.

2. *COOLING TOWER:*
   a. Inspect wiring and electrical controls for loose connections; charring, broken, or wet insulation; evidence of short circuiting and other deficiencies.
   b. Check motor for excessive heat and vibration.
   c. Check electric motor for lubrication.
   d. Check gearbox/drive shaft for lubrication.
   e. Check all piping and connections.
   f. Check fan for bent blades, balance, noise, and vibration.
   g. Inspect mounting brackets, bolts, etc.
   h. Check and inspect belts.
   i. Check pulleys for alignment, as applicable.
   j. Inspect for rust and corrosion. Remove rust and corrosion and apply paint where applicable.
   k. Check nozzles and diffusers.
   l. Check reservoir for leaks and/or missing sealant.
3. *AIR HANDLER, 3–25 TON:*
   a. Check unit for proper operation prior to assessment.

E	U	R	D

b. Turn power off and on.
c. Check access panel.
d. Check cartridge type filters.
e. Check bag type filter.
f. Check electrical wiring and connections.
g. Inspect motor for excessive heat and noise.
h. Check shaft bearings and motor for lubrication.
i. Inspect fan for bent blades and balance.
j. Check belt for wear, proper tension, and alignment.
k. Visually inspect coil.
l. Check coil, drip pan, and blower for cleanliness.
m. Inspect piping and valves for leaks.
n. Check air plenum for cleanliness.
o. Check for corrosion.
p. Check operation of damper motor.
q. Check damper motor for lubrication.
r. Check unit for proper operation after assessment.
s. Check automatic on/off times.

4. *AIR HANDLER, 26–100 TON:*
a. Check air handler for proper operation prior to assessment.
b. Turn power off and on.
c. Check access panel.
d. Check cartridge-type filter.
e. Check bag-type filter.
f. Check electrical wiring and connections.
g. Inspect motor for excessive heat and noise.
h. Check shaft bearings and motor for lubrication.
i. Inspect fan for bent blades or balance.
j. Check belt for wear, proper tension, and alignment.
k. Visually inspect coil.
l. Check coil, drip pan, and fan for cleanliness.

E	U	R	D

m. Inspect piping and valves for leaks.
n. Clean air plenum.
o. Check operation of damper motor.
p. Check damper motor for lubrication.
q. Check unit for proper operation after assessment.
r. Check and adjust automatic off/on timer.

5. *CONDENSER, AIR-COOLED, 3–25 TONS:*
a. Check operation of condenser prior to assessment.
b. Turn power off and on.
c. Check access panel.
d. Check electrical wiring and connections.
e. Inspect motor for excessive noise and heat.
f. Check shaft bearings and motor for lubrication.
g. Inspect fan for bent places or balance.
h. Check belt for wear, proper tension, and alignment.
i. Visually inspect coil.
j. Check coil and fan for cleanliness.
k. Inspect piping and valves for leaks.
l. Check for corrosion.
m. Check operation of damper motor.
n. Check operation of damper motor.
o. Check operation of condenser after assessment.

6. *CONDENSER, AIR-COOLED, 26–100 TONS:*
a. Check operation of condenser prior to assessment.
b. Turn power off and on.
c. Check access panel.
d. Check electrical wiring and connections.
e. Inspect motor for excessive noise and heating.
f. Check shaft bearings and motor for lubrication.
g. Inspect fan for bent blades and balance.
h. Check belt for wear, proper tension, and alignment.

E	U	R	D

i. Visually inspect coil.
j. Check coil and fan cleanliness.
k. Inspect piping and valves for leaks.
l. Check for corrosion.
m. Check operation of damper motor.
n. Check damper motor for lubrication.
o. Check operation of condenser after assessment.

7. *EVAPORATIVE CONDENSERS OR WATER COOLING TOWERS, 3–25 TONS:*
   a. Check operation of unit prior to assessment.
   b. Turn power off and on.
   c. Check access panel.
   d. Check electrical wiring and connections.
   e. Inspect motor for excessive noise and overheating.
   f. Check shaft bearings and motor for lubrication.
   g. Inspect fan for bent blades and balance.
   h. Check belt for wear, proper tension, and alignment.
   i. Check water pump for lubrication and cleanliness.
   j. Inspect prime tube coil.
   k. Check prime tube coil for cleanliness..
   l. Check inside of water tower for cleanliness.
   m. Check water level.
   n. Check float; adjust if necessary.
   o. Inspect piping and valves for leaks.
   p. Check operation of damper motor
   q. Check damper motor for lubrication.
   r. Check unit for corrosion.
   s. Check operation of unit after assessment.
8. EVAPORATIVE CONDENSERS OR WATER COOLING TOWERS, 26–100 TONS:
   a. Check operation of unit prior to assessment.

	E	U	R	D
b. Turn power off and on.				
c. Check access panel.				
d. Check electrical wiring and connections.				
e. Inspect motor for excessive noise and overheating.				
f. Check shaft bearings and motor for lubrication.				
g. Inspect fan for bent blades and balance.				
h. Check belt for wear, proper tension, and alignment if required.				
i. Check water pump for cleanliness and lubrication.				
j. Check spray nozzle.				
k. Inspect prime tube coil for cleanliness.				
l. Check inside of water tower for cleanliness.				
m. Check water level.				
n. Check float.				
o. Inspect piping and valves for leaks.				
p. Check operation of damper motor.				
q. Check damper motor for lubrication.				
r. Visually inspect for corrosion.				
s. Check operation of unit after assessment.				
9. *REFRIGERATION EQUIPMENT:*				
a. Inspect wiring and electrical controls for loose connections; charred, broken, or wet insulation.				
b. Check motor for lubrication.				
c. Check motor for excessive heat and vibration.				
d. Check condition of belt.				
e. Check cooling coils for dust.				
f. Check hermetic unit for housing leaks, evidence of overheating, excessive noise and vibration.				
g. Inspect door gaskets.				
h. Inspect for rust and corrosion.				
10. *AIR COMPRESSORS:*				

	E	U	R	D
a. Inspect wiring and electrical controls for loose connections, charred, broken, or wet insulation, and short circuits.				
b. Check air intake muffler/filter for cleanliness.				
c. Check cylinder fins for cleanliness.				
d. Test all safety valves.				
e. Check oil level in crank case.				
f. Check for condensate in air tank.				
g. Check and inspect belts.				
h. Check foundation bolts for tightness.				
i. Check trigger valve strainers for cleanliness.				
j. Check motor for excessive heat and vibration.				
k. Check electric motor for lubrication.				
l. Check all screws and nuts for tightness.				
11. *AIR COMPRESSOR EQUIPMENT:*				
a. Open and close latch-type lid or cover plate.				
b. Check compressor oil level.				
c. Check tension, condition, and alignment of v-belt on fractional HP compressor.				
d. Check tension, condition, and alignment of v-belts on large compressor.				
e. Drain moisture from air storage tank and check low-pressure cut-in; while draining, check discharge for indication of interior corrosion.				
f. Check air intake filters on compressor for cleanliness.				
g. Check reusable oil filter for cleanliness.				
h. Check operation of pressure relief valve.				
i. Check cylinder cooling fins and air cooler on compressor.				
j. Check bolts, foundation, cylinder head, belt guard, etc.				
k. Check electric motor for excessive heat or vibration and lubrication.				

	E	U	R	D
l. Check compressor power, control circuits, and electrical connections.				
m. Perform operation check of air compressor.				
n. Check fractional HP air compressor for dry sprinkler system. Includes checking belt, oil level, lubrication, and drain condensate.				

Total Items: 11 Raw Total %

Findings: (Comment on each checkpoint and summarize to justify scoring):

____

____

Division V

Fire Protection System Identification

E	U	R	D

I. *OBJECTIVE:* The basic objective is to maintain fire protection equipment in full operating condition at all times. This division does not cover mobile and portable firefighting equipment and appliances primarily associated with organized fire department activities.

II. *DEFINITIONS:* Fire protection equipment covered by the checkpoints includes:

1. Fire alarm panels.
2. Building sprinkler alarm valves.
3. Fire alarm boxes.
4. Security and intrusion alarm systems.

III. *MAINTENANCE STANDARDS.*
Special emphasis is placed upon maintenance of equipment which fully compensates for the reduced amount of personnel needed to detect and combat fire conditions. Maintenance standards for emergency protective equipment shall not be relaxed. Equipment must be maintained in full operative condition.

Section 1. Fire Alarm Panel

E	U	R	D

SCOPE: Installed fire alarm panel boxes.

1. Check system for proper operation.
2. Check batteries and burglar alarm, as required.
3. Inspect for loose or inadequate connections.

Total Items: 3 Raw Total %

Findings (comment on each checkpoint and summarize to justify scoring):

____

____

____

____

____

____

____

____

____

____

____

____

____

E	U	R	D

Section 2. Building Sprinkler Alarm Valves

SCOPE: All sprinkler alarm systems and detection/activation devices and other sprinkler equipment

1. Check fractional horsepower of air compressor (air supply for dry sprinkler or preaction alarm valve systems).
2. Check operational condition of fire department connection (Siamese connection).
3. Call fire alarm control center to place master alarm box out of and into service for sprinkler alarm valve operational check.
4. Reset master alarm box auxiliary trip coil between alarm valve trip tests.
5. Check operation of post indicator valve or outside stem and yoke control valve.
6. Visually inspect sprinkler heads for system.
7. Check general condition of wet alarm valve, alarm valve enclosure, and deluge alarm valve.
8. Check general condition of dry alarm or preaction alarm valve and valve enclosure.
9. Check pressure gauge for accuracy against bench-tested standard gauge.
10. Obtain access to sprinkler valve enclosure or machine room.
11. Perform 2-inch full flow drain test on dry sprinkler deluge sprinkler or preaction sprinkler alarm systems.
12. Perform 2-inch full-flow drain test on wet sprinkler alarm valve system.
13. Perform dry sprinkler alarm valve trip test (includes draining lines and resetting valve).
14. Check operation of post indicator valve and clean indicator windows.

Total Items: 14 Raw Total %

E	U	R	D

Findings (comment on each checkpoint and summarize to justify scoring):

Section 3. Fire Alarm Boxes

SCOPE: Applies to master fire alarm boxes.

1. *GENERAL:*
   - *a.* Check to see that all wiring is tight and in good condition.
   - *b.* Check alarm boxes for water leaks.
   - *c.* Check box ground to ensure that the box ground is not tied to another ground system.
   - *d.* Check boxes and mechanisms for possible lightning damage.
   - *e.* Check terminal connections for possible corrosion.

E	U	R	D

f. Check master boxes to see that they operate on all three-folds.
  1.) Alarm records on metallic system.
  2.) Alarm records on negative wire.
  3.) Alarm records on positive wire.
  4.) Alarm records on ground.

g. Check speed of box.

2. *REMOTE FIRE ALARM BOXES:*
   a. Check to see that all wiring is tight and in good condition.
   b. Check remote box for water leaks.
   c. Check switches for adjustment.
   d. Actuate box and check to see that the remote will actuate the master box.
3. *MUNICIPAL-TYPE MASTER FIRE ALARM BOX:*
   a. Open and close box doors and travel on foot (compact loop) to next box or return to alarm headquarters.
   b. Travel in motorized vehicle from alarm headquarters to first box in an expanded loop.
   c. Open and close box doors and travel in motorized vehicle (expanded loop) to next box or return to alarm headquarters.
   d. Visually inspect interior of box for discrepancies.
   e. Perform operational check through complete cycle and restore box to operational state.
   f. Inspect exterior of box, indicating light and surrounding area.
   g. Test lightning arrestor.
4. *MUNICIPAL-TYPE AUXILIARY CIRCUIT PULL BOXES:*
   a. Travel in motorized vehicle from alarm headquarters to master fire alarm box for auxiliary circuit.
   b. Coordinate tests of circuit with cognizant personnel in protected area. Set local panel for test.

E	U	R	D

c. Travel on foot from custodial area to first pull box in circuit.
d. Open and close pull box with key and travel to the next pull box or return to the custodial area.
e. Visually inspect interior of pull box for a discrepancies.
f. Inspect exterior of pull box.
g. Conduct test of pull box.
h. Set up local equipment for normal operator and notify custodian upon completion of tests.
i. Test lightning arrestor.

5. *FIRE ALARM BOX, LIGHTS:*
   a. Check for cracked or broken luminaries and fixture parts, missing pullcords, and insulating links on pull chains.
   b. Inspect wiring and sockets for loose connectors, broken insulation, and other damage.
   c. Check fire alarm box lights proper location and adequate support.

Total Items: 5 Raw Total %

Findings (comment on each checkpoint and summarize to justify scoring):

____

____

____

____

____

____

E	U	R	D

Section 4. Security and Intrusion Alarm System

SCOPE: Covers security and intrusion alarm system that impact the fire protection system.

1. Check operation of tamper alarm switch (ultrasonic devices, microwave devices, infrared, etc.).
2. Check tamper switch on magnetic door switch.

Total Items: 2 Raw Total %

Findings (comment on each checkpoint and summarize to justify scoring):

Division

VI

Special Systems Assessment

E	U	R	D

I. *OBJECTIVE:* The basic objective is to maintain special systems used for scientific research purposes in an economical manner that will be consistent with functional requirements, sound architectural and engineering practice, and reasonable appearance.

II. *DEFINITIONS:*

1. *FACILITIES:* Special systems include wind tunnels, vacuum spheres/pressure vessels, air-launch/power-check facilities, and rocket-motor test/power-check facilities. Not included are institutional maintenance facilities/items located within or adjacent to the facility. Institutional maintenance facilities/items are covered by checklists promulgated for Category Items Building Systems, Operational Facilities, Utilities and Ground Improvements, Utilities Plants, and Fire Protection Systems.
2. *COMPONENTS:* For purposes of Facilities Condition Assessment, special systems are divided into the following basic components for which assessment checklists are established.

E	U	R	D

a. Structural features (foundations, structures, frames, and similar items).
b. Electrical systems.
c. Plumbing and piping systems.
d. Motor assemblies/electric motors.
e. Fans, fan shafts, and fan shaft bearings.

III. *MAINTENANCE STANDARDS:* Maintenance shall include providing all services and materials required to keep special systems in such a state of preservation that they may be continuously utilized for scientific research purposes.

Section 1. Structural Features (Foundations, Structural Frames, and Similar Items)

SCOPE: Cover foundations for vacuum spheres/pressure vessels, air launch/power-check facilities, and rocket-motor test/power-check facilities; and moorings/decks/blast deflectors on air-launch/power-check facilities and rocket motor test/power-check facilities.

1. Foundation checkpoints as outlined in Buildings Systems Assessment Checklist, Subcategory—Foundations, are to be used for this assessment.
2. Moorings/decks/blast deflectors checkpoints as outlined in Operational Facilities Checklist, Subcategories—Antennas—Supporting Towers and Masts and Fuel Facilities (Receiving and Issues) are to be used for this assessment.

Total Items: 2 Raw Total %

Findings (comment on each checkpoint and summarize to justify scoring):

E	U	R	D

Section 2. Electrical Systems

SCOPE: Covers electrical systems that directly support scientific research activities. Does not cover electrical systems that support institutional maintenance requirements. Checkpoints outlined in Operational Facilities, Subcategories as outlined below, are to be used for this assessment:

1. Disconnecting switches
2. Distribution transformers
3. Electrical instruments
4. Electrical potheads
5. Electric relays
6. Power transformers
7. Electric motors and generators
8. Fuses and small circuit breakers
9. Rectifiers

Total Items: 9 Raw Total %

Findings (comment on each checkpoint and summarize to justify scoring):

E	U	R	D

Section 3. Plumbing and Piping Systems

SCOPE: Covers plumbing and piping systems for Wind Tunnels, Vacuum Spheres/Pressure Vessels, Air-Launch/Power-Check Facilities, and Rocket Motor Test/Power-Check Facilities. Plumbing Systems checkpoints as outlined in Utilities and Ground Improvements Assessment Checklist, Subcategory—Piping Systems, are to be used for this assessment.

Findings (comment on each checkpoint and summarize to justify scoring):

____

____

____

____

____

____

____

Seciton 4. Motor Assemblies/Electric Motors

SCOPE: Covers motor assemblies and electric motors for Wind Tunnels, Vacuum Sphere/Pressure Vessels, Air-Launch/Power-Check Facilities, and Rocket-Motor Test/Power-Check Facilities. Motor assemblies and electric motor checkpoints as outlined in Operational Facilities Assessment

E	U	R	D

Checklist, Subcategory—Electric Motors and Generators, are to be used for assessment.

Findings (comment on each checkpoint and summarize to justify scoring):

____

____

____

____

____

____

____

____

Section 5. Fans, Fan Shafts, and Fanshaft Bearings

SCOPE: Covers fans, fan shafts, and fan shaft bearings for Wind Tunnels.
Fans, fan shafts, and fan shaft bearings checkpoints as outlined in Buildings Systems Assessment Checklist, Subcategory—Ventilating and Exhaust Air Systems, are to be used for this assessment.

Findings (comment on each checkpoint and summarize to justify scoring):

E	U	R	D

Inspection Checklist Summary

Division	Total	Emergency	Urgent	Routine	Deferred
I. Building Systems Assessment					
Chimneys & Stacks	14				
Buildings/Structures	9				
Roofs	17				
Trusses	5				
Trailers	2				
Air Conditioning Syst.	11				
Cranes and Hoists	43				
Elevators/Platform Lifts/Dumbwaiters	8				
Food Preparation and Service Equipment	20				
Heating Systems	4				
Plumbing Systems	21				
Ventilating and Exhaust Air Systems	2				
Hot Water Systems	3				
Electrical Systems	3				
Lighting	7				
Switch Gear	15				
I. SUBTOTAL	**184**				
II. Operation Facilities Assessment					
Tower, Masts, and Antennas	17				
Chemical/Fuel Facilities	20				
Chemical (Fuel Facility Storage)	21				
Brows and Gangways	10				
Camels & Separators	2				
Dolphins	12				

Inspection Checklist Summary

Division	Total	Emergency	Urgent	Routine	Deferred
Piers, Wharves, Quaywalls, & Bulkheads	30				
Disconnecting Switches	5				
Electrical Grounds & Grounding Systems	2				
Electrical Instruments	5				
Electrical Potheads	4				
Electrical Relays	2				
Lightning Arresters	7				
Power Transformers, Deenergized	8				
Power Transformers, Energized	20				
Safety Fencing	11				
Steel Poles/Structures	6				
Vaults and Electrical Manholes	16				
Cathodic Protection Systems	2				
Electrical Motors and Generators	3				
Pier Circuits and Receptacles	8				
Distribution Transformers, Deenergized	4				
Distribution Transformers Energized	12				
Buried & Underground Telephone Cable	2				
Telephone Substation	15				
Fuses & Small Circuit Breakers (600V and below, 30A and below)	8				
Rectifiers	3				

Inspection Checklist Summary

Division	Total	Emergency	Urgent	Routine	Deferred
II. SUBTOTAL	**255**				
III. Utilities and Ground Improvements Assessments					
Bridges and Trestles	26				
Fences and Walls	15				
Grounds	17				
Railroad Trackage	9				
Pavements	5				
Retaining Walls	8				
Storm Drainage System	14				
Tunnels and Underground Structure	15				
Piping Systems	12				
Steam Distrib. Equip.	17				
Pumps	4				
Fresh Water Supply & Distribution System	23				
Sewage Collection & Disposal Systems	13				
Unfired Pressure Vessels	4				
Underground Tanks	16				
III. SUBTOTAL	**198**				
IV. Utilities Plants					
Fresh Water Supply	2				
Electrical Generator and/or Distribution	10				
Heating Generation	9				
Air-Conditioning Generation & Distribution	11				

Part

3

Assessment Inspection Checklists (Abbreviated)

Part 3 is an abbreviated checklist as compared with the more detailed checklist shown in Part 2. Space is provided in the condition column, to the left of each checklist item. Check off whether that particular item is considered *good, fair, or poor* (where "G" = *good*, "F" = *fair, and* "P" = *poor*). At the end of each section, divide the number of *good, fair, or poor* by the total number of checklist items to obtain a percentage of that condition.

BUILDING SYSTEMS ASSESSMENT

BLDG.: ______________________________

INSPECTOR: ______________________________

DATE: ______________________________

CONDITION			ELECTRICAL/PLUMBING
G	F	P	
___	___	___	1. What is the condition of outlets, switches, breakers, and alignment of contacts, and are there any signs of arcing?

G	F	P	
___	___	___	2. Is there any indication of loose connections or broken wires on emergency lights?
___	___	___	3. How frequently are pilot/power lights, transformer, and rectifier tested on emergency lighting systems?
___	___	___	4. What is the condition of lead-line conductors and lightning arrestors?
___	___	___	5. Is there any indication of chips or foreign object damage on lightning arrestors?
___	___	___	6. What is the condition of motor-pump assemblies?
___	___	___	7. Is there any indication of leaks, low oil levels, excessive noise and vibration, rust and corrosion, excessive dirt in filter and strainer assembly in motor-pump assemblies?
___	___	___	8. What is the condition of the motor-pump assembly and relief valve operation?
___	___	___	9. Is there any indication of fluid circulation problems, excessive noise and vibration, pump volume control cylinder leaks, electrical equipment and connections short circuiting, loose pipe hangers and supports, and rust or corrosion on hydraulic motor-pump assemblies?
___	___	___	10. What is the condition of all connections, circuit breaker contacts, oil levels, valves and gaskets on oil circuit breakers?
___	___	___	11. What is the condition of all wiring and electrical control connections on panel boards? Is there any indication of rust or corrosion?
___	___	___	12. What is the condition of piping, valves, and fixtures? Is there any indication of rust, corrosion, leaking, scale, clogging, or other damage?

G	F	P	
—	—	—	13. What is the indication of piping insulation? Is there any indication of asbestos, open seams, missing sections/fastenings, or other damage?
—	—	—	14. What is the condition of platforms, pedestals, and supports? Is there any indication of rust, corrosion, missing/loose connections, defective parts, misalignment, or other damage?
—	—	—	15. What is the condition of steam traps, strainers, and bypass valves? Is there any indication of leakage or other damage?
—	—	—	16. What is the condition of wiring/electrical controls, electric motor, fan shaft bearings, bolts/brackets, wall switch/thermostat, water/steam/gas lines, and/or steam traps and burners on unit heaters? Is there any indication of loose connections, short circuits, or leaks, etc.?
—	—	—	17. What is the condition of thermal insulation/protective coverings, burner assemblies, electrical heating elements and controls on water heaters? Is there any indication of loose, damaged, or missing parts, loose connections, short circuits, rust, or corrosion?
—	—	—	18. Is there any indication of loose, rusted/corroded cable or cable racks?
—	—	—	19. What is the condition of the elevator motor assembly, doors, tracks, and cables? Is there any indication of rust/corrosion, dust, grease, loose connections/broken wires on any of the electrical elements of the elevator equipment?
—	—	—	TOTAL

SCORE:

\# *GOOD*

______________________ = %

TOTAL NO. ITEMS

\# *FAIR*

______________________ = %

TOTAL NO. ITEMS

\# *POOR*

______________________ = %

TOTAL NO. ITEMS

Findings (comment on each checkpoint and summarize to justify scoring):

____

____

____

____

____

____

____

____

____

____

____

____

____

____

BLDG.: __________________________________

INSPECTOR: ______________________________

DATE: __________________________________

CONDITION			EXTERIOR ELEMENTS
G	F	P	
__	__	__	1. Is there any indication of spalls, breaks, exposed reinforcement, settlement of concrete surfaces?
__	__	__	2. Is there any indication of eroded joints, spalling bricks, settlement/buckling of any of the masonry surfaces?
__	__	__	3. Is there any indication of stains, holes, rust/corrosion on any of the exterior wall surfaces?
__	__	__	4. Is there any indication of holes, rust/corroded, or non-fitting door/window screens?
__	__	__	5. Is there any indication of loose, missing sealant, rust/corroded, or rotted door/window frames?
__	__	__	6. Is there any indication of misaligned, rusted/corroded, clogged, leaking, missing guards/fasteners on any gutters/downspouts?

G	F	P	
___	___	___	7. What is the condition of all tile, slate, asphalt-roll, asphalt-shingle, built-up roofing? Is there any indication of excessive weathering, broken, cracked, loose, missing, alligatoring, buckling, blistering, insufficient lapping, tearing, low spots and/or ponding of water, failure of or lack of gravel stops, cracks in membrane, exposed bituminous coatings, exposed, disintegrated, blistered, or buckled felts?
___	___	___	8. What is the condition of single-ply membrane (both modified bitumin and/or EPDM) roofing? Is there any indication of insufficient lapping of membranes, tearing, cracking, low spots and ponding of water, failure or the lack of ballast stops, cracks in exposed membrane where you have ballasted roofs?
___	___	___	9. What is the condition of standing-seam metal roofing? Is there any indication of holes, looseness, punctures, broken seams, inadequate side and endlaps, inadequate expansion joints, rust or corrosion?
___	___	___	10. What is the condition of fasteners, cap and vertical flashing? Is there any indication of rust, missing vertical joints/flanges, improper fastening, improper sealing, missing cants, buckling, cracking, or failed surface coats?
___	___	___	11. What is the condition of chimney, wall, ridge, vent, valley, and edge flashing? Is there any indication of open joints, loose, improper fastenings, and/or other damage?
___	___	___	12. Is any indication of cracks, spalling, missing expansion joints, low spots, or poor drainage?

G	F	P	
__	__	__	13. What is the condition of parapet walls and copings? Is there any indication of cracks, spalling, joint deterioration, or other damage?
__	__	__	14. Is there any indication of rust/corrosion at power-operated doors, loose connections at electrical controls, broken/wet insulation, loose chain drives/belts, or door tracks and bearings?
__	__	__	TOTAL

SCORE:

\# *GOOD*

______________________ = %

TOTAL NO. ITEMS

\# *FAIR*

______________________ = %

TOTAL NO. ITEMS

\# *POOR*

______________________ = %

TOTAL NO. ITEMS

Findings (comment on each checkpoint and summarize to justify scoring):

____

____

____

____

____

____

____

____

____

____

____

____

____

____

____

BLDG.: ______________________________________

INSPECTOR: __________________________________

DATE: _______________________________________

CONDITION	HEATING, VENTILATION, AND AIR CONDITIONING
G F P	
__ __ __	1. How frequently are the air conditioning unit filters changed and what are the conditions of the coils?

G	F	P	
__	__	__	2. Is there any indication of leaks, loose wiring and electrical control connections, frayed/broken insulation, or any rust and corrosion?
__	__	__	3. What is condition of belts, shaft bearings, motors, wiring electrical controls, insulation, exhaust system ducts, collectors, smoke pipes, and hoods, and is there any indication of short circuiting, and any presence of rust or corrosion?
__	__	__	4. What is the condition of belts, belt tension, air filters, wiring and electrical controls on heating/ventilating units?
__	__	__	5. Is there any indication of loose connections, charred, frayed or broken insulation, rust or corrosion on heating and ventilating units?
__	__	__	6. Is there any indication of loose wiring and electrical control connections, broken insulation, excessive heat/vibration, hermetic housing leaks, missing or loose door gaskets, rust or corrosion on refrigeration equipment?
__	__	__	7. What is the condition of wiring, electrical controls, electric motors, bolts/brackets, switches, thermostatic controls, water/steam/gas lines, steam traps or boiler burners in central heating plants? Is there any indication of loose connections, short circuits, or leaks in any of the systems described above?
__	__	__	8. What is the condition of air handling units, compressors, absorption units, wiring, electrical controls, electric motors, bolts, shafts, fan shaft bearings, switches, thermostatic controls, water/steam lines in central air conditioning plants? Is there any indication of loose connections, short circuits, or leaks in any of the systems described above?

G	F	P	
__	__	__	TOTAL

SCORE:

# *GOOD*
______________________ = %
TOTAL NO. ITEMS

# *FAIR*
______________________ = %
TOTAL NO. ITEMS

# *POOR*
______________________ = %
TOTAL NO. ITEMS

Findings (comment on each checkpoint and summarize to justify scoring):

____

____

____

____

____

____

____

____

BLDG.: ______________________________

INSPECTOR: ______________________________

DATE: ______________________________

CONDITION	CLEAN ROOMS
G F P	
__ __ __	Are all clean room filter systems changed as frequently as required? Is there any indication of loose cable connections on the fan motors or other exhaust system components?

Findings (comment on each checkpoint and summarize to justify scoring):

____

____

BLDG.: ________________________________

INSPECTOR: ____________________________

DATE: _________________________________

CONDITION			INTERIOR FINISHES
G	F	P	
__	__	__	1. Is there any indication of curling. loose, fading, wrinkling, or stained interior wall coverings?
__	__	__	2. Is there any indication of worn, chipped, torn, raveled, stained, or unbonded floor coverings?
__	__	__	3. Is there any indication of checking, blistering, scaling, peeling, wrinkling, fading, mildewed, or unbonded painted surfaces?
__	__	__	4. Is there any indication of loose, missing, warped, stained, or faded ceiling tile?
__	__	__	TOTAL

SCORE:

\# *GOOD*

____________________ = %

TOTAL NO. ITEMS

\# *FAIR*

____________________ = %

TOTAL NO. ITEMS

POOR

_________________________ = %

TOTAL NO. ITEMS

Findings (comment on each checkpoint and summarize to justify scoring):

____

____

____

____

____

____

____

____

____

____

BLDG.: ____

INSPECTOR: ____

DATE: ____

CONDITION			STRUCTURAL
G	F	P	
___	___	___	1. Is there any indication of warping, sagging, deflection, rotting, damaged, or missing bolts in any of the wood timber/framing?
___	___	___	2. Is there any indication of sagging, warping, rotting, or stained floors or stairs?
___	___	___	TOTAL

SCORE:

GOOD

________________________ = %

TOTAL NO. ITEMS

FAIR

________________________ = %

TOTAL NO. ITEMS

POOR

________________________ = %

TOTAL NO. ITEMS

Findings (comment on each checkpoint and summarize to justify scoring):

____

____

____

____

____

____

____

____

____

____

____

____

____

EXTERIOR FACILITIES ASSESSMENT

BLDG.: ________________________________

INSPECTOR: ____________________________

DATE: _________________________________

CONDITION			ELECTRICAL/PLUMBING
G	F	P	
__	__	__	1. Is there any indication of rust/corrosion, dust, dirt, grease on electrical terminals, cables, grounds, and/or continuity?
__	__	__	2. Is there any indication of loose, missing, cracked, rotted, or insect-infested wood members on wood towers?
__	__	__	3. What is the condition of luminaries, fixtures, lenses, insulation, etc.?
__	__	__	4. Is there any indication of loose connections, charred/broken, wet insulation, or evidence of short circuiting of any of the wiring/electrical controls?

G	F	P	
—	—	—	5. Is there any evidence of damaged wiring devices or defective insulators?
—	—	—	6. Is there any evidence of excessive cable sag/spacing, or number of conductors in conduits/raceways?
—	—	—	7. What is the condition of exposed and/or underground piping? Is there any indication of leakage, corrosion, loose connections, or other damage?
—	—	—	8. What is the condition of exposed and/or underground valves? Is there any indication of bent stems, leakage, or corrosion?
—	—	—	9. What is the condition of meters? Is there any indication of leakage, corrosion, broken glass, or moisture behind the glass?
—	—	—	10. What is the condition of hydrants and hydrant shut-off valves? Is there any indication of missing caps, broken/missing chains, damaged threads, guards, I.D. markings, or rust or corrosion?
—	—	—	11. What is the condition of check valves, meter pit manholes, roadway boxes, manhole frames/covers/ladders/rungs? Is there any indication of rust, corrosion, or other damage?
—	—	—	12. What is the condition of walkways, guardrails, stairs, and ladders? Is there any indication of rust, corrosion, broken or missing parts?
—	—	—	13. Is there any indication of rust or corrosion, broken or frayed cable, or loose connections at cable terminals and lead jumpers on cathodic protection systems?
—	—	—	14. Is there any indication of rust or corroded anode suspensions; bent/broken suspension members/braces, frayed/broken suspension cables, or loose bolts or cable connections on cathodic protection systems?

G	F	P	
___	___	___	15. Is there any indication of loose connections, broken/wet insulation at wiring/ electrical controls on cathodic protection systems?
___	___	___	16. What is the condition of engine oil, fuel, radiator and/or battery water on emergency generators?
___	___	___	17. What is the condition of all electrical wiring, connections, switches, brushes, contacts on emergency generators? Is there any indication of rust or corrosion?
___	___	___	18. What is the condition of electrical switchgear? Is there any indication of overheating and looseness of connections, rust or corrosion?
___	___	___	19. What is the condition of telephone lines? Is there any indication of improper clearances over private/public property, waterways, streets and sidewalks, from electric light/power lines, trees, contact wires/ transformers, and/or improper sag or debris on wire?
___	___	___	20. What is the condition of bridle cables/ wires? Is there any indication of loose connections, abraded insulation, kinks, uninsulated splices, proper placement/termination?
___	___	___	21. What is the condition of connections at cable terminals, binding posts, bridging and/ or test connectors?
___	___	___	22. What is the condition of transformer bushings and insulators? Is there any indication of rust, corrosion and/or arcing?
___	___	___	23. What is the condition of transformer wiring and electrical controls? Is there any indication of loose connections, broken/wet insulation, evidence of short circuiting, or loose mountings?

G	F	P	
—	—	—	24. What is the condition of transformer foundations and supporting pads? Is there any indication of movement, damage, rot, or insect infestation?
—	—	—	25. What is the condition of transformer grills and louvers? Is there any indication of improper ventilation, presence of moisture, active desiccant and weather tightness?
—	—	—	TOTAL

SCORE:

GOOD

____________________ = %

TOTAL NO. ITEMS

FAIR

____________________ = %

TOTAL NO. ITEMS

POOR

____________________ = %

TOTAL NO. ITEMS

Findings (comment on each checkpoint and summarize to justify scoring):

____

____

____

____

____

____

____

____

____

____

____

____

____

BLDG.: ______________________________________

INSPECTOR: __________________________________

DATE: _______________________________________

CONDITION			GROUNDS
G	F	P	
___	___	___	1. What is the condition of all turf areas? Is there any indication of traffic damage, discoloration, bare spots, weeds, diseases, insect damage, erosion, or excessive turf height?
___	___	___	2. What is the condition of trees/shrubs? Is there any indication of interference with utilities, buildings, recent structural/storm damage, diseases, or insect damage?
___	___	___	3. What is the condition of border areas? Is there any indication of poison/noxious weeds, seedling trees, erosion and siltation, lack of vigor, inadequacy of coverage, or evidence of burning?

G	F	P	
__	__	__	4. What is the condition of earthen dams/dykes? Is there any indication of damage from erosion, burrowing animals, seepage, lack of vegetation, logs, debris at outlet ends, pipe damage, or failure?
__	__	__	5. What is the condition of emergency spillways? Is there any indication of blockage or erosion damage?
__	__	__	6. What is the condition of cut slopes and diversion channels? Is there any indication of erosion, scour, burning, weakness from past overflow, lack of sufficient vegetation coverage, or inadequate surface runoff piping?
__	__	__	7. What is the condition of sprinkler system nozzles, sprays, hose, pipe, and valves? Is there any indication of rust, corrosion, clogging, inadequate pressures, or leakage?
__	__	__	8. What is the condition of catch basins and curb inlets? Is there any indication of debris, obstructions, cracked, broken, or improperly seated grating, or settlement?
__	__	__	9. What is the condition of pipelines? Is there any indication of settlement, cracked, broken or open joints, sediment, debris, or tree roots, erosion in concrete pipes, erosion/corrosion of corrugated metal pipes?
__	__	__	10. What is the condition of concrete headwalls? Is there any indication of cracked, broken, spalling, exposed reinforcing, settlement, and undermining? Verify condition of pipe joint at headwall.
__	__	__	11. What is the condition of concrete foundation/retaining walls and concrete or masonry retaining walls? Is there any indication of spalling, cracking, or other forms of deterioration?

G	F	P	
___	___	___	12. What is the condition of timber retaining walls? Is there any indication of rot, insect infestation, or other forms of deterioration?
___	___	___	13. What is the condition of embankment slopes and areas behind walls? Is there any indication of excessive erosion or settlement?
___	___	___	14. What is the condition of manhole frames, covers, and ladder rungs? Is there any indication of rust, corrosion, missing or broken parts?
___	___	___	15. What is the condition of manhole walls? Is there any indication of cracking, spalling, exposed reinforcing, eroded/sandy mortar joints, loose, broken, or displaced brick?
___	___	___	16. What is the condition of manhole bottoms? Is there any indication of clogging flow and excessive silt?
___	___	___	17. What is the condition of culverts? Is there any indication of sediment or obstructions at inlets/outlets, ditch bottoms not flush with pipe inverts, and channeling?
___	___	___	18. What is the condition of gutters and ditches? Is there any indication of cracked, broken, eroded concrete surfaces, expansion joint alignment obstructions, ponding of water, or inadequate side vegetation coverage?
___	___	___	TOTAL

SCORE:

\# *GOOD*

____________________ = %

TOTAL NO. ITEMS

\# *FAIR*

______________________ = %

TOTAL NO. ITEMS

\# *POOR*

______________________ = %

TOTAL NO. ITEMS

Findings (comment on each checkpoint and summarize to justify scoring):

____

____

____

____

____

BLDG.: ______________________________________

INSPECTOR: __________________________________

DATE: _______________________________________

CONDITION			PAVEMENTS
G	F	P	
—	—	—	1. What is condition of curbs, gutters, expansion joints/filler? Is there any indication of cracks, breaks, alignment problems, damaged tops, or loose/missing/unbonded joint filler?
—	—	—	2. What is the condition of rigid pavement? Is there any indication of spalling, cracking, depressions, and/or buckling?

G	F	P	
___	___	___	3. What is the condition of flexible pavement? Is there any indication of spalling, cracking, depressions, and/or buckling?
___	___	___	4. What is the condition of brick/stone pavement? Is there any indication of loose/missing pieces, or bedding/grout failure?
___	___	___	5. What is the condition of gravel, cinder, shell, and stabilized soil for breaks? Is there any indication of potholes, deterioration, or other damage?
___	___	___	6. What is the condition of airfield taxi/runway pavements? Is there any indication of spalling, cracking, depressions, and/or buckling of pavements?
___	___	___	TOTAL

SCORE:

GOOD

_________________________ = %

TOTAL NO. ITEMS

FAIR

_________________________ = %

TOTAL NO. ITEMS

POOR

_________________________ = %

TOTAL NO. ITEMS

Findings (comment on each checkpoint and summarize to justify scoring):

____

____

____

____

____

BLDG.: ________________________________

INSPECTOR: ____________________________

DATE: _________________________________

CONDITION			RAILROAD TRACKAGE
G	F	P	
__	__	__	1. What is the condition of all trackage? Is there any indication of vertical/horizontal misalignment, or sinking where track passes from earth fill to bridges?
__	__	__	2. What is the condition of track rails? Is there any indication of breaks, splits, cracks, loose or missing bolts at joints, splices, etc.?
__	__	__	3. What is the condition of signal lamps? Is there any indication of loose connections, insufficient battery current, rust, or corrosion?
__	__	__	4. What is the condition of railroad ties? Is there any indication of decay, splitting, deterioration, improperly imbedded in ballast, or inadequate drainage?
__	__	__	5. What is the condition of railroad crossings? Is there any indication of roughness to traffic, and/or obstructions?
__	__	__	6. What is the condition of ballast? Is there any indication of excessive dirt/mud accumulations, soft/wet spots, erosion, or settlement?

G	F	P	
__	__	__	7. What is the condition of warning signs? Is there any indication of inadequate placement, illegible and/or unstable bumper blocks, and/or damaged guardrails?
__	__	__	TOTAL

SCORE:

\# *GOOD*

______________________ = %

TOTAL NO. ITEMS

\# *FAIR*

______________________ = %

TOTAL NO. ITEMS

\# *POOR*

______________________ = %

TOTAL NO. ITEMS

Findings (comment on each checkpoint and summarize to justify scoring):

____

____

____

____

____

BLDG.: ______________________________

INSPECTOR: ______________________________

DATE: ______________________________

CONDITION			STRUCTURAL
G	F	P	
__	__	__	1. Is there any indication of rust/corrosion, loose steel or aluminum members on gangways or platforms?
__	__	__	2. Are there conditions of rust/corrosion on anchor bolts, splices/rivets, straps, structural steel members, ladders, etc. on antennae support towers/masts?
__	__	__	3. Is there any indication of rust/corrosion, cracking, scaling, peeling on painted support tower surfaces?
__	__	__	4. Is there any indication of cracked/broken, spalled concrete or exposed reinforcement on antennae support tower/masts concrete foundations?
__	__	__	5. What is the condition of steel members? Is there any indication of twist, sag, rupture, shearing/crushing of steel plates, members bolts, rivets, loose bolts and rivets, or broken welds?
__	__	__	6. What is the condition of protective coatings? Is there any indication of blistering, checking, cracking, scaling, flaking, rust, or corrosion?
__	__	__	7. What is the condition of steel power pole concrete bases, pads and anchor bolts? Is there any indication of cracks, breaks, settlement, rust, corrosion, loose or missing bolts?
__	__	__	8. What is the condition of poles, structures, crossarms, and beams? Is there any indication of rust, corrosion, misalignment, and loose bolts?

G	F	P	
__	__	__	9. What is the condition of power pole guys, anchors, and ground wires? Is there any indication of rust, corrosion, damaged hardware/anchors, missing or damaged insulators, corroded guy shields, and excessive sag or tension?
__	__	__	10. What is the condition of wood poles, crossarms, and buckarms? Is there any indication of decay, damage, splits, alignment, and/or insect infestation?
__	__	__	11. What is the condition of insulators, pins, guys, grounds, ties, and/or line wires? Is there any indication of looseness, breaks, cracks, rust, and corrosion?
__	__	__	12. What is the condition of vault manhole covers, gratings, vault doors, ladders, lights, switches, cables, potheads, and junction boxes? Is there any indication of defective gaskets, cracks, rust, corrosion, damaged hinges, locks, latches, missing/broken rungs, and/or missing parts?
__	__	__	TOTAL

SCORE:

\# *GOOD*

________________________ = %

TOTAL NO. ITEMS

\# *FAIR*

________________________ = %

TOTAL NO. ITEMS

\# *POOR*

________________________ = %

TOTAL NO. ITEMS

Findings (comment on each checkpoint and summarize to justify scoring):

____

____

SPECIAL SYSTEMS

BLDG.: ______________________________________

INSPECTOR: __________________________________

DATE: _______________________________________

CONDITION			WIND TUNNELS
G	F	P	
__	__	__	1. What is the condition of all wiring, electrical controls, motor assemblies, fans, fan shafts, and fan shaft bearings? Is there any indication of loose connections, short circuits, etc.?

Findings (comment on each checkpoint and summarize to justify scoring):

____

____

____

____

____

BLDG.: ______________________________

INSPECTOR: ______________________________

DATE: ______________________________

CONDITION VACUUM SPHERES/PRESSURE VESSELS

G	F	P	
__	__	__	2. What is the condition of vacuum spheres/pressure vessels wiring, electrical controls, electric motors, compressors, valves, bolts, brackets, switches, etc.? Is there any indication of loose connections, short circuits, or leaks?

Findings (comment on each checkpoint and summarize to justify scoring):

____

____

____

____

____

____

____

____

____

____

BLDG.: ______________________________

INSPECTOR: ______________________________

DATE: ______________________________

CONDITION			TUNNELS/UNDERGROUND STRUCTURES
G	F	P	
__	__	__	3. What is the condition of tunnel/underground structure? Is there any indication of improper drainage, cracks, breaks, leaks, eroded slopes, missing/damaged wing walls, erosion of slopes, loose rocks, missing door and gate operating/locking devices, rust, corrosion, loose, missing, or other damaged parts?
__	__	__	4. What is the condition of tunnel concrete floors? Is there any indication of cracks, breaks, or other damage; proper grading and drainage for earth and gravel floors; leaks, settlement, breaks, or other damage at tunnel linings; and spalling, disintegrating, or loose rocks at unlined tunnels?
__	__	__	5. What is the condition of tunnel metal roofs, pipelines, and pipeline supports/anchors? Is there any indication of rust, corrosion, broken/missing parts, and adequate supports?
__	__	__	6. What is the condition of tunnel ventilation equipment, grounding connections, and lighting systems? Is there any indication of improper operation, rust, corrosion, loose, missing, or damaged parts?

Findings (comment on each checkpoint and summarize to justify scoring):

____

____

____

____

____

____

____

____

BLDG.: __________________________________

INSPECTOR: ______________________________

DATE: __________________________________

CONDITION			BRIDGES
G	F	P	
__	__	__	7. Is there any missing, broken, decayed or eroded bridge foundation riprap cribbing, bulkheads, and/or piles?
__	__	__	8. Are there cracked, broken, and/or disintegrated concrete/bituminous surfaces, and/or curb and gutters at roadway approaches?
__	__	__	9. Is there any indication of excessive settlement at joint between fill and structure?
__	__	__	10. Are there any missing or damaged bridge railings, load and speed limit signs?
__	__	__	11. Is there any indication of cracked, scaling, exposed concrete reinforcement at all bridge concrete foundations?

G	F	P	
__	__	__	12. Is there any indication of cracks, breaks, spalling, or disintegrated/open joints; impact and/or vibration damage at concrete bridge abutments and piers?
__	__	__	13. Is there any indication of rust, corrosion, loose, missing, bowed, bent, or broken steel members?
__	__	__	14. Is there any indication of rusted, corroded, missing or loose bridge seats; rusted, corroded, or unlubricated bridge rollers; frayed, raveled, broken, or defectively anchored bridge cable?
__	__	__	15. Is there any indication of rusted, corroded, loose, missing, broken welds, or other damage to bridge splices, bolts, rivets, screws, etc.?
__	__	__	16. Is there any indication of rusted, corroded, cracked, scaled, peeled, wrinkled paint on bridge surfaces?

Findings (comment on each checkpoint and summarize to justify scoring):

____

____

____

____

____

BLDG.: ______________________________

INSPECTOR: ______________________________

DATE: ______________________________

CONDITION			FUEL/CHEMICAL STORAGE/ DISTRIBUTION FACILITIES
G	F	P	
__	__	__	17. Is there any indication of leaks, corrosion, damaged or missing hangars/supports, or other deficiencies to fuel pipes, check valves, meters and gages?
__	__	__	18. Is there any indication of leaks, obstructions, excessive damage, or wear to fuel strainers or shock arrestors?
__	__	__	19. Is there any indication of damaged or clogged vent screens?
__	__	__	20. Is there any indication of loose connections, lack of electrical continuity, rust or corrosion on fuel systems electrical grounding systems?
__	__	__	21. Is there any indication of damaged, missing/broken, rusted/corroded ladder rungs, platforms, and walkways?
__	__	__	22. Is there any indication of missing, damaged, or illegible signs?
__	__	__	23. Is there any indication of loose/missing fuel island platform planks, damaged or deteriorated framing supports, guardrails and stairs?
__	__	__	24. Is there any indication of rusted/corroded metal hose racks and reels?
__	__	__	25. Is there any indication of a loss of, cracked, scaled, or wrinkled painted surfaces?
__	__	__	26. Is there any indication of cracked, spalling, or exposed concrete reinforcement to fuel tank foundations?

G	F	P	
___	___	___	27. Is there any indication of rusted/corroded or deteriorated exterior painted steel tank, structural support surfaces?
___	___	___	28. Is there any indication of damaged, rusted, corroded expansion-type roofs, seals, supports, and support-guides?
___	___	___	29. Is there any indication of damaged tank linings?
___	___	___	30. Is there any indication of cracked, broken, spalling, rusted/corroded; with any settling, heaving, water seepage, with adequate sod covering on outer face at dikes?

Findings (comment on each checkpoint and summarize to justify scoring):

____

____

____

____

____

____

____

____

____

____

BLDG.: ____________________

INSPECTOR: ____________________

DATE: ____________________

CONDITION			PIERS
G	F	P	
__	__	__	31. Is there any indication of missing, broken/loose connections, obstructions of curbings, handrails, and catwalks?
__	__	__	32. Is there any indication of cracks, holes, or other damage to asphalt deck coverings?
__	__	__	33. Is there any indication of rust, corrosion, broken, bent or missing ladder rungs or rotted deck planking?
__	__	__	34. Is there any indication of broken, decayed, or termite-infested wood stringers, pile caps, and/or bearing piles?
__	__	__	TOTAL

SCORE:

# *GOOD*
____________________ = %
TOTAL NO. ITEMS

# *FAIR*
____________________ = %
TOTAL NO. ITEMS

# *POOR*
____________________ = %
TOTAL NO. ITEMS

Findings (comment on each checkpoint and summarize to justify scoring):

Part

4

Supplementary Assessment Inspection Checklists

Part four consists of supplementary checklists of a more general nature. Areas that are briefly examined are:

- Safety management
- Custodial management
- Grounds management
- Environmental management
- Indoor air quality management
- Predictive maintenance
- Energy management program

SECTION 1. SAFETY MANAGEMENT

SCOPE: Safety management provides a system to constantly monitor the work environment to minimize potential safety and health threats.

1. GENERAL SAFETY:
   a. Is safety mentioned in the organization's mission statement?
   b. Is there consistent enforcement of safety policies and procedures?
   c. Has input been considered from all affected areas: office areas, loading dock, production areas, distribution areas, mail room, supply and storage areas, laboratory areas, mechanical rooms?
   d. Has a standardized accident investigation report form been developed and used?
   e. Has a safety program been developed and implemented?
      - Have safety policies been established?
      - Have safety and health committees been established?
      - Are safety inspections conducted by managers?
      - Is there a safety training program?
      - Is there an initial safety orientation for new employees?
      - Have on-the-job and off-the-job safety programs been initiated?
      - Is there a safety, reward incentive program?
2. WORKPLACE SAFETY:
   a. Personal Protective Equipment (PPE):
      - Do personnel have hearing protection equipment?
      - Do personnel have clothing protection equipment?
      - Do personnel have face, eye and foot protection equipment?
      - Is there a self contained breathing apparatus available for personnel to use?
   b. Machinery protection and safeguards:
      - Are portable hand tools properly used?
      - Are equipment belts, chains, gears, and moving parts checked, aligned, lightened, and replaced as required?
      - Is preventive and predictive maintenance performed properly? Are motors and bearings checked, cleaned, and properly lubricated?

c. Environmental safety:
   - Is lighting adequate?
   - Is the facility ventilation and air quality adequate?
     - Are there complaints of musty odors?
     - Are exhaust fans cleaned periodically?
     - Are filters changed on a scheduled basis?
   - Are stained ceiling tiles replaced within a few days?
   - Are leaking pipes/coils repaired quickly?
   - Do doors and locksets operate properly?
   - Are work stations clean, trash removed?
   - Are restrooms clean?
   - Is there a solid waste/recycling program?
   - Are gas cylinders stored properly?
   - Are chemicals (cleaning supplies, paint, etc.) stored properly?
   - Are Material Safety Data Sheets (MSDS) immediately available?
   - Are hallways/corridors free from obstructions?
   - Are floors sagging, cracked, blistering, or slippery?

d. Electrical safety:
   - Do electrical outlets work properly?
   - Are electrical outlets grounded?
   - Are extension cords properly used?

e. Fire protection:
   - Are fire exit signs visible and at known locations?
   - Are smoke detectors available?
   - Are fire extinguishers checked as required?
   - Is there a fire alarm system? Are there manual pull stations?
   - Is there a sprinkler system?

3. CONFINED SPACE:

a. Is there a confined space procedure?

b. Is training in confined space provided? Topics should include:
   - Knowing the hazards
   - Recognizing signs and symptoms of exposure to the hazard
   - Understanding the consequences and risks

- Maintaining contact with a qualified person
- Notifying the qualified person when self-initiated evacuation occurs
- Proper use of personal protective equipment
- Lockout/tagout procedures
- Emergency extraction

c. Is the air in the confined space tested for dangerous air contaminants or oxygen deficiency?
d. Prior to entering a confined space, is the space atmosphere checked, flushed with air, or purged of flammable, incapacitating substances?
e. Is care used not to use oxygen-depleting equipment or materials which will lower the oxygen level below 19.5%?
f. Are mechanical devices, such as blowers, used to force air movement in the space?
g. How often is monitoring performed (should be every 15 minutes)?
h. Are employees required to wear Personal Protective Equipment (mask, self-contained breathing apparatus, hard hat, hard-toed shoes, insulated coveralls, etc.)?
i. Are Material Safety Data Sheets reviewed? Are they readily available?
j. Is the confined space area, if located in public space, secured with barricades?
k. Is mechanical personnel extraction equipment available in the event of an emergency?
l. Is there a Confined Space Attendant appointed?
m. Is the Confined Space Attendant positioned at the entry point of the confined space?
n. Is there a means of communication available, onsite, in the event an emergency situation occurs?
o. Has the Confined Space Attendant been trained on the hazard and emergency extraction?

4. LOCKOUT/TAGOUT:
   a. Before servicing or maintaining a piece of machinery or equipment is that machinery/equipment turned off or disconnected from the energy source?

- Is the energy isolating device locked or tagged out?
- Are lockout devices being used affixed in a manner that ensures the energy isolating device is "safe" or "off"?
- Is the use of lockout devices recorded in a daily log, specifying mechanic's name, date and time installed, and location?

b. Is there an effective lockout/tagout training program established?
   - Is retraining conducted?
   - Is proof available that can show which employee attended?
   - Does training cover the following topics?
     - Type and magnitude of hazardous energy sources?
     - Means of isolating and controlling energy sources?
     - Recognizing when control procedures should be implemented?
     - What to do when a lockout device is in place?

c. Do lockout procedures include notifying all affected employees?

d. Are tagout devices used when it is not possible to isolate/
   render inoperative the machinery or equipment to be serviced?
   - Are tagout devices properly approved? Non-reusable, attachable by hand, self-locking, with minimum unlocking strength of no less than 50 pounds?
   - Is the use of tagout devices recorded in a daily log specifying mechanic's name, date and time installed, location and purpose?
   - Does the tagout device include the following: "Do not start, Do not open, Do not close, Do not energize, Do not operate!"?

e. Who is authorized to remove tagout devices?

f. Is stored or residual energy (e.g., capacitors, springs, flywheels, hydraulic systems, and air, gas or water pressure, etc.) relieved, dissipated, or restrained by accepted methods such as: grounding, bleeding down, blocking, etc.?

g. Are contractors informed of the organization's lockout/tagout procedures prior to commencing work?
h. Are annual inspections of lockout/tagout procedures conducted?

Findings (comment on each checkpoint and summarize to justify scoring):

____

____

____

____

____

____

SECTION 2. CUSTODIAL MANAGEMENT

SCOPE: Custodial management is unique with every organization. This is the one major function that employees and visitors see immediately, and it influences their perception of the facility.

1. MANAGEMENT:
   a. Is there one person in-charge overall of custodial cleaning?
   b. Is there a qualified supervisor on each shift?
   c. Is there an inspection program in place?
   d. Do supervisors spend most of their time, in the field, meeting with workers?
   e. Are Material Data Safety Sheets (MSDS) available on all supplies and chemicals?
   f. Has a quality assurance program been implemented?
   g. Are all complaints followed up?
   h. Are workers informed of inspection results?

i. Is there a training program?
j. Are new workers given an orientation and information folder?
k. Are custodians encouraged to obtain certification?
l. Is project work planned and scheduled?
m. Is there a measurement system in place?
n. Is attendance monitored?
o. Are in-house cleaning services periodically compared with contract cleaning services (costs? quality? productivity?)?
p. Are uniforms provided for employees?
q. Is there an employee suggestion program?
r. Are workers provided lockers?
s. Is there a procedures manual? Does every custodian have a copy?
t. Is absenteeism less than 4%?

2. EQUIPMENT:
   a. Is there sufficient equipment available for use?
   b. Is the equipment the correct type and size for the work being done?
   c. Have employees been trained on how to operate and maintain the equipment?
   d. Is the equipment permanently marked with an inventory control number?
   e. Is equipment/tools secure from theft, vandalism?
   f. Are radios and cell phones available for supervisors to use?
   g. Is there a preventive and predictive maintenance program in place for all custodial equipment?

3. SUPPLIES:
   a. Are adequate amounts of cleaning supplies available for employees?
   b. Have employees been trained how to use the supplies and the impacts of mixing chemicals?
   c. Are there proper controls on the purchasing of supplies?
   d. Do custodians mix their own cleaning supplies?
   e. Have personnel been educated on the use of Material Safety Data Sheets?

f. Is there an inventory control system in place? Is bar coding used?
g. Are annual purchase agreements in place?
h. Are products and materials standardized?
i. Are all cleaning supply containers properly marked?
j. Are cleaning supplies delivered to workers in their facility?
k. Are expendable cleaning supplies (mops, sponges, wiping cloths, etc.) easily available to workers?

4. CLEANING ISSUES:
   a. Are cleaning procedures standardized?
   b. Are employees knowledgeable as to what type of equipment should be used for specific work?
   c. Is the preponderance of cleaning done at night?
   d. Are custodial closets properly sized and equipped?
   e. Are time standards for cleaning established?
   f. Are square footage standards for cleaning established?
   g. Are personnel trained on how to respond to emergencies?
   h. Are custodians used for non-cleaning activities?
      - Furniture moving
      - Event setups
      - Snow removal
      - Running errands
   i. Are custodians used to move solid waste (trash) and recycling?

5. METHODS OF CLEANING:
   a. Area assignment:
      - Is all cleaning work done by an individual on a repetitive basis, in one geographic area?
      - Is good work recognized?
      - Is competition encouraged?
   b. Specific job cleaning:
      - Is only one specific cleaning function performed by a custodian?
   c. Team cleaning:
      - Is a cleaning team (2–3 workers) used for project work?

d. Gang cleaning:
   - Does the entire cleaning crew clean and move from space to space?
   - Is the work speed of the crew continuously monitored?

Findings (comment on each checkpoint and summarize to justify scoring):

____

____

____

____

____

____

____

SECTION 3. GROUNDS MANAGEMENT

SCOPE: Grounds management is important because of the first impression conveyed to employees and visitors to a facility.

1. PLANNING AND SCHEDULING:
   a. Is a computer maintenance management system (CMMS) used to plan and schedule work?
   b. Is there a plant materials maintenance program which specifies schedules for grass cutting, tree trimming and pruning, pesticide application, etc.?
   c. Is there an inventory of all plant materials?
   d. Is there a litter control plan?
   e. Has an effective integrated pest control plan been implemented?

   f. Are chemicals (pesticides/herbicides) properly stored?
   g. Has a shrub/plant winterization program been developed?
   h. Does the surrounding landscape have good drainage?
   i. Are sufficient trash and recycling receptacles available on the grounds?
   j. Are regular fire safety inspections (leaves, flammable storage) of the grounds made periodically.
   k. Is grounds maintenance scheduled in order to minimize noise interference to others?
   l. Has a grounds and landscape site plan been developed? Is it kept current?
   m. Is there a snow removal plan?

2. EQUIPMENT:
   a. Has a grounds equipment and vehicle preventive maintenance program been implemented?
   b. Is there sufficient grounds equipment on hand?
      - Is it in good operating condition?
      - Is it adequate for the need?
   c. Are automatic sprinkler systems onsite?
      - Are they centrally controlled?
      - Is periodic maintenance performed on the sprinkler heads, controllers?
   d. Is there an equipment replacement program?
   e. Are personnel issued seasonal weather clothing?
   f. Is equipment kept clean, free of rust spots, and painted?

3. TRAINING:
   a. Is there a grounds training program?
   b. Is training conducted during inclement weather?
   c. Does training cover minimum federal, state, and local requirements?
   d. Does training emphasize safety and OSHA requirements?
   e. Are grounds personnel properly trained on the operation and maintenance of their equipment?
   f. Is all training documented?
   g. Are personnel properly trained and certified on pesticide application?

h. Have personnel been trained on how to read Material Safety Data Sheets?
i. Have personnel been trained on recognizing incorrect irrigation methods?
j. Has a Hazard Communications Standard Program been developed?
    - Have personnel been trained on it?

4. GROUNDS MAINTENANCE:
    a. Turf maintenance:
        - Drainage of turf is very important. Is the slope of the turf area less than the bare minimum of 1%?
        - Is turf area aerated annually?
        - Is fertilizer selected based on the existing soil condition?
        - Are areas that have erosion repaired quickly?
            - Are areas around catch basins covered with grass?
            - Are slopes covered with protective vegetation?
            - Is mulch placed around trees where grass will not grow?
            - Are building rain downleaders connected to the storm water system?
            - Are gullies along walks and drives filled in quickly?
            - Is mulch used to protect against erosion?
        - Is grass cut to specified height? (Note: excess clippings can cause disease and fungus problems.)
        - Are grass clippings removed?
        - Are weeds kept under control?
        - Is turf thatch removed mechanically? (Note: removal of thatch is necessary. For turf to thrive disease-free it needs water and air.)
    b. Water management:
        - Is there a turf and plant irrigation system available?
        - Does the irrigation system have an automatic controller and rain sensor allowing the system to turn on and off independently?
        - Is there a centralized computer system to control watering times based on climate, weather conditions, soil type and compaction, turf variety, and landscape features?

    - Can this system operate other features such as fountains, gates, lighting, and security systems?
    - Is there a sewer measuring meter connected to the system? (Note: this device measures the amount of water savings on sewer charges.)
  - Is this system connected to the municipality water system or a well?
  - Is there a method in place to monitor fungus growth, thatch accumulation, and soil compaction which can result from irrigation?
    - Is there a spraying and aeration program in place to keep the above in check?
  - Has the proper moisture level for various plants and shrubs been determined?

c. Plant and shrub maintenance:
  - Are plants and shrubs selected for the right size space?
  - Is there overcrowding of the space?
  - Is soil acidity checked periodically?
  - Is the area monitored for proper drainage?
  - In selecting plants, is consideration given to:
    - Geographic location?
    - Local conditions?
  - Is the hole where the plant is to be placed dug to the correct depth? (Note: it should be no deeper than the depth, and no wider than the ball to be planted. Backfill with topsoil and soil conditioner.)
  - Has a fertilization program been developed for the cultivated conditions and various types of plants?
  - Has a plant pest control program been developed?
    - Are organic pesticides used?
  - Is pruning done to:
    - Remove dead or damaged branches?
    - Control size of plant?
    - Retain original shape?
    - Control disease of infected branches?
    - Remove suckers?
  - Are hedges pruned wider at the bottom than the top?
  - Are shrub beds free of tree seedlings?

d. Indoor landscaping:
- Are foliaged plants used in public space (entrances, hallways, lobbies, lounges)?
  - Are they located near a light source (preferably natural sunlight)?
  - Are plants exposed to artificial light 12–16 hours per day? (Note: fluorescent is better than incandescent.)
- Does the temperature in the area where plants are located range between 55 and 80°F?
- If there is low humidity, are plants placed on trays of moist pebbles?
- Is watering of foliated plants done when the top 1 or 2 inches of soil is dry?
  - Is water added from above?
  - Is lukewarm water being used?
- Is there an increase in the amount of fertilization (10–30%) for plants grown in artificial light as compared to those grown in natural light?
- Is fertilization done every three to six months?
- Is falling foliage cleaned up regularly?
- Are plants periodically cleaned of dust?

e. Other functions:
- Policing of grounds:
  - Is paper and debris removed from grounds on a daily basis?
- Graffitti:
  - Is graffitti removed when detected?
- Paved surfaces:
  - Are potholes repaired?
  - Are storm water obstructions removed?
  - Are spalling, broken concrete surfaces repaired or replaced?
  - Are uneven stones, missing brick, or depressions repaired?
  - Are vehicle wheel stops straight and secured to pavement?
  - Is grass near walks and roadways edged?
- Storm drainage:

    - Are catch basins cleaned out annually?
    - Are storm drains cleaned of leaves, snow, and ice?
  - Physical features:
    - Are signs and light poles straightened and painted?
    - Are fences repaired and painted?
    - Are trash containers clean and setting level?
    - Are outside benches and picnic tables clean, repaired, and painted?
    - Are bicycle racks neat and orderly?
    - Are statues and sculptures periodically waxed?
    - Are loading docks free of trash?
    - Are bulletin boards and kiosks straightened, repaired, and painted?
    - Are planters weeded and mulched?
    - Is outdoor lighting in working order?
    - Is grass neatly trimmed around trees, structures, fences, poles, etc.?

5. SOLID WASTE AND RECYCLING:
   a. Solid waste:
      - Has a waste flow analysis been conducted?
      - Is solid waste hauled to a nearby landfill?
      - Are private haulers used to deliver solid waste for disposal?
      - Has the organization developed a waste disposal system?
      - Have waste requirements been identified?
        - Type of waste?
        - Quantity of waste?
      - Is waste disposal handled in-house?
      - Are compactors maintained by in-house mechanics?
      - Are solid waste containers cleaned, serviced, and repaired on a preventive maintenance basis?
      - Is rust on waste containers removed and then the containers painted?
      - Is waste disposed of by incineration?
   b. Recycling:
      - Is recycling mandated by the local municipality?
      - Has a waste composition analysis been conducted?
      - Do custodians haul recycling materials from the facility to the trash location site?

- Are provisions made to recycle:
  - Paper (office, mixed, news)?
  - Glass?
  - Aluminum?
  - Scrap metal?
  - Toner cartridges?
  - Oil, glycol?
  - Yard waste?
  - Construction rubble?
  - Cardboard?
  - Computers?

Findings (comment on each checkpoint and summarize to justify scoring):

____

____

____

____

____

____

____

____

____

SECTION 4. ENVIRONMENTAL MANAGEMENT

SCOPE: This section applies to a variety of management challenges involving the maintenance, disturbance, abatement and worker safety in all facilities.

1. ASBESTOS ABATEMENT:
   a. Has an Asbestos Abatement Manager been appointed?
   b. Has an Asbestos Management Plan been developed? Does it contain the following:
      - Historical background
      - Method of abatement: in-house or contract
      - Summary of asbestos assessment:
        - Location specified
        - Condition of asbestos:
          - Loose
          - Friable: more than 1% by weight which can be crumbled/pulverized by hand
          - Solid
      - What steps did the organization decide to take in order to comply with EPA regulations?
        - Complete removal
        - Enclosing the asbestos
        - Encapsulating in order to seal in the asbestos
   c. Has an independent off-site laboratory been contracted with to conduct independent testing for asbestos?
   d. Are proper sampling techniques used (note: preferably by a qualified laboratory)?
      - Is there a chain of custody record?
      - Have photographs been taken of sample sites?
      - Have sample sites been described thoroughly?
        - Description
        - Exact location
        - Date/time collected
        - Investigator's name
   e. Are proper analytical methods used (note: preferably by a qualified laboratory)?
   f. Have signs been placed at locations containing asbestos, alerting personnel of the presence of asbestos and the potential for exposure?
   g. Have Asbestos Abatement files been established?
      - Historical record of each abatement
      - Air quality reports of each abatement
      - Record of procedure for each abatement

- Disposal record
- Kept for at least 20 years

h. Has a public relations program been developed and put in place?

i. Has action been taken to prevent asbestos fibers from being released into the air? (Note: OSHA specifies no employee should be exposed to airborne asbestos in concentrations greater than 0.02 fiber per cubic centimeter without personal protective equipment.)
- Taping
- Sealing
- Wrapping
- Wetting

j. Is periodic air monitoring conducted?

k. Are air monitoring results recorded and filed?

l. Are qualified contractors used for demolition?

m. Is asbestos containing waste properly disposed?
- No visible emissions discharged?
- Treatment with water?
- Acceptable asbestos disposal site?

n. Is there an asbestos training and awareness program?
- Have personnel been reminded that asbestos containing materials are not dangerous as long as they are not disturbed, causing fibers to become airborne?
- Are personnel aware of the various types of asbestos?

2. LEAD MANAGEMENT:

a. Has a Lead Abatement Manager been appointed?

b. Are contractor's required to comply with the organization's Lead Management program?

c. Have facilities been inspected for lead-based paint and other sources of lead?

d. Once lead-based paint is identified, is it kept in good repair, i.e., intact with no peeling or chipping?

e. Have water sources been tested for lead, specifically, the plumbing system: pipes, fittings, type of solder used, and faucet fixtures?

f. To reduce the potential for lead found in drinking water, have the following steps been followed?

- Let the tap water run for several seconds in order to flush any lead that could be settled in the faucet.
- Install a domestic water pipe charcoal filter.
- Replace faucet known to contain lead.
- Make use of bottled water for drinking and cooking.
- Replace piping, fittings, and solder known to contain lead.
- Check with the local municipality to have water service line to the facility checked for lead content.
- Check electrical wiring for grounding conditions which may increase the possibility of corrosion.

g. Is there a lead safety training and awareness program?
   - Are maintenance personnel and custodians required to attend?
   - How often is lead awareness training conducted?

h. Are trained, qualified personnel/contractors used to conduct lead testing?
   - Have they been certified in accordance with the EPA Lead Accreditation Program?

i. Is Personal Protective Equipment (PPE) available for workers who have to work in lead-contaminated areas?

j. Has a Lead Abatement Plan been developed? Does it contain the following:
   - Location of the lead abatement project
   - Dates work was done
   - Approximate amount of lead-containing material removed
   - Method used for abatement
   - Medical surveillance provision for workers
   - Disposal method
   - Documentation showing certification of the individual or company performing the abatement
   - Permit from local municipality
   - Verification that abatement worker safety training was conducted

k. During an abatement project, is there effective site containment?

l. Have custodians been trained on safe practices to clean dust and debris which will result from the abatement process?

- Is detergent and wet washing used?
- Are carpets vacuumed with HEPA filtered vacuums?
- Are floors coated and sealed?

m. Has the organization decided on the abatement method to be used?
   - Removal and replacement
   - Enclosing the lead by sealing
   - Encapsulating

n. Are records kept of all abatement projects? (Note: Records must be kept a minimum of three years.) Do they include the following:
   - Location?
   - Name and address of individual who supervised the project?
   - The amount of lead-based paint removed?
   - Dates and times when the abatement took place?
   - Name and address of disposal sites?

Findings (comment on each checkpoint and summarize to justify scoring):

____

____

____

____

____

____

____

____

____

____

SECTION 5. INDOOR AIR QUALITY MANAGEMENT

SCOPE: Indoor Air Quality (IAQ) Management is important because 90% of the workday, for most employees, is spent indoors. Recent estimates indicate that exposure to pollution is two to five times higher indoors than outdoors. Trying to determine if there is an air quality problem, and its cause, is very difficult.

1. CAUSES OF INDOOR AIR QUALITY (IAQ) PROBLEMS:
   a. Inadequate ventilation:
      - Is sufficient fresh outdoor air entering the building? (See ASHRAE Standard for recommended limits of acceptable indoor air volume.)
      - Is there adequate air distribution and mixing which can minimize stratification, draftiness, and pressure differences within the building?
      - Do temperature and humidity levels fluctuate excessively?
      - Is the building ventilation system properly maintained?
        - Are filters changed on a regular basis?
      - Is the need to conserve energy (reducing or eliminating fresh outdoor air; lowering thermostats or economizer cycles in winter and raising them in summer; reducing/eliminating humidification or dehumidification, etc.) causing building ventilation problems?
      - Have building exhaust and intake vents been installed properly; location, height?
      - What does the HVAC operations and maintenance profile show of the building in question?
   b. Poor filtration:
      - Is the organization using high-quality, efficient filters?
      - Are filters properly sized and fit?
      - Are filters properly installed in order to reduce the amount of unfiltered air which could impact heat transfer at the coils?

c. Mechanical system cleanliness:
   - Is there stagnant water inside air-handling systems?
   - Are air handling unit coils clean?
   - Is all mechanical system water being chemically treated?

2. SYMPTOMS AND SIGNS:
   a. Are building occupants complaining of: headaches, dizziness, eye or skin irritation, shortness of breath, nausea, fatigue, sinuses, coughing and sneezing?
   b. Are more than 20% of the building occupants showing the above symptoms?
   c. Do symptoms continuously persist? Do they disappear during non-work hours?

3. CAUSATIVE AGENTS:
   a. Physical:
      - Temperature
      - Humidity
      - Condensation
   b. Chemical:
      - Pesticides
      - Carbon dioxide (CO_2)
      - Carbon monoxide (CO)
      - Off-gasing
      - Tobacco smoke
      - Volatile organic compounds
      - Particulates
   c. Microbiological:
      - Bacteria
      - Mold
      - Fungi
   d. Other sources:
      - Bird droppings
      - Insects
      - Algae
      - Mites

4. IAQ INSPECTION TECHNIQUES:
   a. Analysis of building data:

- Have technical data on mechanical systems been obtained by examining building mechanical plans?
- Have Operations and Maintenance manuals for the various mechanical systems in the building been checked?
- Are system air balancing and testing reports available?
- Is information available on:
  - Type of air handling unit, size, operational characteristics, etc.?
  - Exhaust system and fan operation?
  - Type of heating and cooling system?
  - Type of humidification system and control details?

b. HVAC inspection:
- Has visual inspection been made of: interior of the air handling unit chamber, fan, coils, condensate piping and drain pan, humidifiers and controls?
- Has the filtration system been checked for cleanliness?
- Have static pressure measurements, inside ductwork, been taken at various locations?
- Is cooling tower water chemically treated to prevent microbial contamination?
- Are condensate pans and humidifier reservoirs kept clean of waterborne microbials?

c. Air quality testing:
- Have airborne microbial samples been taken and sent to an approved testing laboratory?
- Is the amount of particulates in the air, indoor and outdoor, being measured?
- Are temperature and humidity levels recorded periodically?

5. INSPECTION TOOLS:

a. Borescope
b. Infrared camera
c. Psychrometer
d. Velometer and thermal anemometer
e. Air sampling pumps

f. Particle counter
g. Thermometer
h. Infrared gas spectrometer
i. Smoke tubes

6. IAQ PREVENTION:
   a. Operations:
      - Is sufficient outside air brought into the building? (Note: Guideline can be found in ASHRAE Standard 62-1989.)
      - Is the building maintained at a positive pressure in order to prevent unconditional/unfiltered outside air from infiltrating into the building?
      - Are building mechanical systems operating suitably due to space renovations? Should these systems be retrofitted?
      - Are Variable Air Volume (VAV) boxes allowed not to close fully in order to always introduce minimum outside air?
      - Can space temperature be lowered, which will increase relative humidity and decrease the potential of occupant susceptibility to indoor contaminants?
        - Relative Humidity (RH) should range between 30% and 60%. (Note: higher RH can result in mold and fungus. Lower RH can result in discomfort and drying of mucous membranes.)
      - Are supply air diffusers free of obstructions?
   b. Maintenance:
      - Are ventilation rates documented on a yearly basis? Does the documentation include:
        - Random space ventilation testing?
        - Temperature readings?
        - Humidity levels?
        - Carbon dioxide testing?
      - Are scheduled preventive maintenance checks and balances conducted?
      - Are all area and exhaust fans operating as designed and is the building properly pressurized?
      - Are restroom floor drains inspected at frequent intervals (weekly) and water or mineral oil added to the

drain trap? (Note: floor drain traps tend to dry out through evaporation and allow sewer gas to infiltrate into the space.)
- Are restrooms kept at negative pressure?
- Are all mechanical system air handlers inspected monthly for overall cleanliness, proper operation of mechanical components, condition of filters, cleanliness and drainage of condensate pans?
- Are all supply air diffuser air volumes checked, at a minimum, semiannually?
- Is carpeting shampooed on a continuous basis?
- Are floor care activities (pollutant generators), such as buffing, sweeping, and vacuuming, conducted when the building is least occupied?

c. Design:
- Do building designs comply with local indoor air standards?
- Are air intake and exhausts located properly and not near sources of outdoor pollution, such as dumpster sites, loading docks, cooling towers, etc.?
- Are ducts and pipes insulated so there are no breaks where surfaces are exposed to air and condensation can occur?
- Do ventilation rates comply with applicable codes?
- Are access doors available to reach every system component?

7. REPORTS/RECORDS:

a. Reports. Should contain:
- Executive summary
- Summary of the present HVAC system
- Building system measurements (temperature, relative humidity, airflow, percent outside air)
- Latest test results:
  - Type of test
  - How conducted
  - Method used
  - Procedure
  - Results

    - Comparison with previous tests
  - Photographs
  - Recommendations
- b. Records:
  - Retain all air sampling data for 40 years.

Findings (comment on each checkpoint and summarize to justify scoring):

SECTION 6. PREDICTIVE MAINTENANCE

SCOPE: Monitoring equipment condition at appropriate intervals to enable accurate evaluation to use as input when determining whether maintenance action or "no action" is required without sacrificing equipment reliability.

1. Has vibration monitoring been used to identify the frequencies of vibration of machines and equipment?
2. Has infrared (IR) thermography been used to identify equipment faults by increases in temperature?
3. Has a formal lubricant analysis program been established to identify equipment lubrication: condition, need, and wear?
4. Has ultrasound been used to detect the high-frequency, low-level sound generated by the early stages of improperly lubricated bearings, leaking valves, and other faults that appear in the frequency range above 20,000 Hz?
5. Have dedicated, motivated personnel with the authority to perform the required functions been appointed? Are they held accountable for program results?
6. Has critical equipment necessary for continued operation of process been selected for the program?
7. Has the proper monitoring frequency been established for each piece of machinery?
8. Have alert levels been established for each piece of critical equipment to provide a flag or trigger for additional necessary actions?

Findings (comment on each checkpoint and summarize to justify scoring):

____

____

____

____

____

____

SECTION 7. ENERGY MANAGEMENT PROGRAM

SCOPE: Maintaining comfort conditions within building spaces while at the same time managing energy costs. This entails both the mechanical systems and the control systems that regulate the mechanical systems.

1. Has there been a baseline energy audit conducted of the facility to have a reference point for comparison to determine actual savings? This comparison will provide an energy usage per degree day for both the heating and cooling mode of the facility.
2. Based on the baseline energy usage audit, have plans been developed and implemented to decrease total energy usage?
3. If energy usage costs are higher than the Building Owners and Managers Association (BOMA) benchmarking guidelines, has the outer envelope of all buildings been reviewed for thermal conductivity; and have the mechanical systems been reviewed for ventilation, exhaust air removed from buildings, the type of fan systems and control system strategy, and mechanical efficiency of various components used.
4. Have you determined that air handling unit controls are working properly?
5. Has the sequencing between the outside, exhaust, and return air mixing dampers been checked?
6. Have control sequences been formulated so that there is no overlap between heating and cooling sequences?
7. Has advantage been taken of microprocessor technology to utilize low-cost variable-frequency drives to vary the speed and frequency of fans?

8. Are pneumatic control systems utilized in larger buildings where long-term stable control systems and distribution of power via pneumatic tubing is an economic alternative?
9. Communications levels have different protocols and may operate at different speeds. Have the following protocols been considered:
   - Intrapanel communications?
   - Sublevel communications?
   - Peer-to-peer level communications?
   - System-wide (global) communications?
10. Has use of the following listed energy management programs been evaluated:
    - Time-of-day programming?
    - Optimized start and stop?
    - Warm-up and cool-down?
11. Has economizer operation been implemented to save energy?
12. Have load shedding and duty cycling systems use been considered?
13. Have reset programs use been considered?
14. Have all energy-savings possibilities within the chiller plant been considered?
15. Have cooling towers been retrofitted with variable frequency drives to more exactly control the fan speed of the condenser water setpoint?
16. Are all air-handling units networked to a command control station to determine the valve position in each unit?
17. Has the hot water supply temperature been reset based on the load being served?
18. Has the energy management control system been programmed in a building management system to turn lights on and off during occupied and unoccupied time periods?

Findings (comment on each checkpoint and summarize to justify scoring):

Appendix

1

Suggested Average Useful Life of Facility Components

	Life Expectancy (years)
I. Major Construction—Primary Structure	
A. Foundation	
1. Concrete block	100
2. Cement	50
3. Waterproofing	
a. Bituminous coating	10
b. Pargeting	20–30
c. Termite treatment	5
B. Exterior walls	
1. Reinforced concrete frame	
a. Masonry exterior	
1) Heavy	45
2) Light & medium	40
2. Steel frame	
a. Masonry exterior	
1) Heavy	45

	Life Expectancy (years)
2) Medium	35
3) Light	30
b. Metal exterior	
1) Heavy	30
2) Medium	25
3) Light	20
3. Wood frame	
a. Masonry exterior	
1) Heavy	35
2) Medium	25
b. Metal exterior	
1) Heavy	30
2) Medium	25
3) Light	20
c. Wood Exterior	
1) Heavy	25
2) Light & medium	20
C. Floors	
1. Wood	35
2. Concrete	45
3. Metal	40
D. Roof	
1. Structure	
a. Wood	35+
b. Concrete	45+
c. Metal	40+
2. Covering	
a. Built up (depends on materials/drainage)	12–25
b. Rubber	15
c. Elastomeric	10
d. Metal (depends on gauge/quality)	20–50+
e. Asphalt	15–30
f. Tile	30–50

	Life Expectancy (years)
g. Slate	50–100
h. Wood shingles/shakes	15–40
II. Secondary Structure	
A. Ceilings	
1. Plaster	lifetime
2. Gypsum board	35+
3. Acoustic tile	15+
B. Interior partitions	
1. Plaster	30–70
2. Gypsum board	30–70
3. Wood	35+
4. Masonry	lifetime
C. Windows	
1. Metal frame	15–20
2. Wood frame	40–50
3. Screen	25–50
4. Storm windows	lifetime
D. Doors	
1. Metal exterior	15–25
2. Wood exterior	15–25
3. Metal frame exterior	15–25
4. Wood frame exterior	15–25
5. Wood interior	30-life
6. Screen	25–50
7. Folding	30-life
8. Garage doors	20–50
9. Garage door opener	15–25
E. Floors	
1. Oak/pine	lifetime
2. Slate flagstone	lifetime
3. Vinyl sheet or tile	20–30
4. Terrazzo	lifetime
5. Carpeting (depends on quality and traffic)	5–10
6. Marble	lifetime

	Life Expectancy (years)
III. Electrical and Mechanical Service Equipment	
A. Electrical systems	
1. Lighting systems	
a. Conduit	lifetime
b. Fixtures	15+
c. Wiring (copper clad or plated)	100+
d. Flood lighting	15
2. Power feed wiring	
a. Bus Duct	25
b. Capacitor	20
c. Power feed wiring mains	25
d. Switchboards	20
e. Switch units	20
3. Transformers	
a. Wet type	20
b. Dry type	15
B. HVAC systems	
1. Air conditioning systems	
a. Central, including ducts & piping	15
b. Window type	10
c. Cooling towers	15
d. Air conditioning compressors	15
e. Rooftop air conditioners	15
2. Heating systems	
a. Furnaces & boilers	20–30
b. Radiators, convectors, piping	25
c. Heat pumps	15
d. Unit heaters (gas or electric)	15–20
3. Humidifiers	8–10
4. Water heaters	
a. Electric	14
b. Gas	10–15
5. Induction fan coil units	20

	Life Expectancy (years)
6. Dampers	20
7. Condensate pans	20–25
8. Coils	
a. DX, water, or steam	20
b. Electric	15
9. Heat exchangers	20–25
10. Burners	20
11. Ventilating systems including fans & exhaust	15
C. Plumbing systems	
1. Drinking water systems	15
2. Fixtures	
a. Enamel steel sinks	25–30
b. Enamel cast iron sinks	25–30
c. China sinks	25–30
3. Piping	
a. Cast iron waste	lifetime
b. Concrete	20–30
c. Copper	lifetime
d. Plastic	30–40
e. Steel	20
f. Vitrified tile	30
4. Faucets (depends on finish/ quality)	15–20
5. Sprinkler systems	
a. Wet & dry systems	30
b. Fire pumps	20
1. Hose housing	
a. Wood	15
b. Steel	20
c. Masonry	30
6. Sump pumps	
a. Small	10
b. Large	15
D. Service systems	

	Life Expectancy (years)
1. Elevators (all types)	20
2. Fire alarm	20
3. Intercom	15
4. Telephone	15
IV. Miscellaneous Items	
A. Bulkheads	
1. Concrete	30
2. Steel	25
3. Timber	20
B. Chimneys	
1. Brick or concrete	35
2. Steel—lined	25
3. Steel—unlined	20
C. Culverts	
1. Concrete	35
2. Galvanized steel	20
D. Curbing	
1. Concrete	25
E. Fencing	
1. Brick or stone	30
2. Chain link	20
3. Concrete	30
4. Wire	10
5. Wood	10
F. Flagpoles	25
G. Incinerators	
1. Commercial type, steel fire brick lined	20
2. Concrete block or brick	20
3. Steel	15
H. Paving and walks	
1. Asphalt on gravel or stone	15
2. Brick	20
3. Concrete	20
4. Gravel, stone, cinders	10

	Life Expectancy (years)
5. Parking area guard rails	10
I. Platforms	
1. Reinforced concrete	35
2. Wood frame concrete piers	20
3. Wood frame on wood posts	15
J. Railroad sidings	25
K. Reservoirs, concrete	35
L. Retaining walls	
1. Brick	30
2. Concrete	40
3. Steel	25
4. Stone	40
5. Wood	15
M. Sheds	
1. Brick, tile, or concrete block with wood frame	25
2. Brick, tile, or concrete block with steel frame	35
3. Metal clad, steel frame	25
4. Metal clad, wood frame	25
5. Wood siding and frame	20
V. Appliances	
A. Compactors	10
B. Dishwashers	10
C. Dryers	14
D. Disposal	10
E. Freezers	
1. Compact	12
2. Standard	16
F. Microwave ovens	11
G. Ranges	
1. Free-standing and built-in, electric	17
2. Free-standing and built-in, gas	19
3. High oven, gas	14

	Life Expectancy (years)
H. Refrigerators	
1. Compact	14
2. Standard	17
I. Washers (automatic and compact)	13
J. Exhaust fan	20
K. Bathrooms	
1. Cast iron bathtub	50
2. Fiberglass bathtub and shower	10–15
3. Shower doors (medium quality)	25
4. Toilet	50
L. Cabinetry	
1. Kitchen cabinets	15–30
2. Bath vanities/wall cabinets	15–20
M. Countertops	
1. Laminate	10–15
2. Ceramic Tile	lifetime
3. Wood/butcher block	20–25
4. Stone/granite	20+
N. Home security appliances	
1. Intrusion systems	15
2. Smoke detectors	10–15
3. Smoke/fire intrusion systems	10
O. Landscaping	
1. Wooden decks	15
2. Brick/concrete patios	25
3. Tennis courts	10
4. Asphalt driveways	10
5. Swimming pool	15–20
6. Turf sprinkler system	10–15
7. Fences	10–15
P. Paints and stains	
1. Exterior paint (wood, metal, masonry)	5–10
2. Interior	
a. Wall paint	5–10

	Life Expectancy (years)
b. Trim and door	5–10
c. Wallpaper	7
Q. Siding	
1. Gutters and downleaders	30
2. Siding	
a. Wood (100 years if properly maintained)	10–100
b. Metal (steel)	50+
c. Aluminum	20–50
d. Vinyl	50

Appendix 2

Electrical Reference Sources

There are many reference sources that a plant engineer can use for electrical projects. Below is a list of some current standards, practices, and guidelines and their source. This list is not intended to be a complete list and there are also many books written on the subject.

1. American National Standards Institute (ANSI)
   Headquarters
   1819 L Street, NW
   Washington, DC 20036
   Telephone: (202) 293-8020

   ANSI Administrative Operations and domestic and international standards facilitation programs remain in New York.
   11 West 42nd Street
   New York, NY 10036
   Telephone: (212) 642-4900
   Website: *www.ansi.org*
   Website: *www.nssn.org* is an online card catalog for technical and standards information.

   ANSI coordinates and administers this country's private sector voluntary standardization system.

2. The Institute of Electrical and Electronics Engineers (IEEE)

445 Hoes Lane
P.O. Box 1331
Piscataway, NJ 08855-1331
Telephone: 1-800-678-4333
Website: *www.ieee.org*

IEEE has a large number of standards that apply to electrical equipment found in industrial and commercial facilities.

- IEEE Color Book Series:
- IEEE-141 (Red Book). Recommended practice for Electric Power Distribution for industrial plants.
- IEEE-142 (Green Book). Recommended practice for grounding of industrial and commercial power systems.
- IEEE-241 (Gray Book). Recommended practice for electric power systems in commercial buildings.
- IEEE-242 (Buff Book). Recommended practice for protection and coordination of industrial and commercial power systems.
- IEEE-399 (Brown Book). Recommended practice for industrial and commercial systems analysis.
- IEEE-446 (Orange Book). Recommended practice for emergency and standby power systems for industrial and commercial applications.
- IEEE-493 (Gold Book). Recommended practice for the design of reliable industrial and commercial power systems.
- IEEE-602 (White Book). Recommended practice for electric systems in health care facilities.
- IEEE-739 (Bronze Book). Recommended practice for energy conservation and cost-effective planning in industrial facilities.
- IEEE-902 (Yellow Book). Guide for maintenance, operation and safety of industrial and commercial power systems.
- IEEE-1015 (Blue Book). Application guide for low-voltage circuit breakers used in industrial and commercial power systems.
- IEEE-1100 (Emerald Book). Recommended practice for powering and grounding sensitive electronic equipment.

3. Illuminating Engineering Society of North America (IESNA)
   120 Wall Street, 17th Floor

New York, NY 10005
Telephone: (212) 248-5000
Website: *www.iesna.org*

The Illuminating Engineering Society of North America establishes and disseminates lighting recommendations to all parties interested. Its Lighting Handbook is recognized as the most authoritative lighting reference in the lighting field. In addition to the handbook the society publishes recommended practices, design guides, lighting memorandums, and technical memorandums.

4. Insulated Cable Engineers Association (ICEA)
   P.O. Box 440
   South Yarmouth, MA 02664
   Telephone: (508) 394-4424
   Website: *www.icea.net*

   The ICEA develops wire and cable standards for the telecommunications and electric utility industries.
   Standards:
   - ICEA Power Cable Standards
   - ICEA Communication Standards
   - ICEA Power Cable Standards published by the National Electrical Manufacturers Association.

5. National Electrical Manufacturers Association (NEMA)
   1300 North 17th Street
   Suite 1847
   Rosslyn, VA 22209
   Telephone: (703) 841-3200
   Website: *www.nema.org*

   The National Electrical Manufacturers Association develops standards for the electrical manufacturing industry. The association publishes over 200 standards.

6. National Fire Protection Association (NFPA)
   1 Batterymarch Park
   P.O. Box 9101
   Quincy, MA 02269-9101
   Telephone: (617) 770-3000
   Website: *www.nfpa.org*

The National Fire Protection Association provides fire, electrical, and life safety to the public. Below is a partial list of current codes and standards that are associated with electrical safety.

NFPA 70: National Electrical Code
NFPA 72: National Fire Alarm Code
NFPA 73: Residential Electrical Maintenance Code for One and Two Family Dwellings
NFPA 75: Protection of Electronic Computer/ Data Processing Equipment
NFPA 77: Recommended Practice on Static Electricity
NFPA 101: Life Safety Code
NFPA 110: Standard for Emergency and Standby Power Systems
NFPA 111: Stored Energy Emergency and Standby Power Systems

There are many additional NFPA standards that are specific to different industries. NFPA also publishes various handbooks that expand on the information contained in the standards. Several of these are: The *NEC Handbook, Fire Alarm Signaling Systems Handbook,* and the *Electrical Installations in Hazardous Locations.*

Appendix

3

Electrical Formula and Data

Volt (*V*) is the unit of electric pressure or electromotive force. It is the potential which will produce a current of 1 ampere through a resistance of 1 ohm.

Ampere (*A*) is the unit of electrical current (coulombs per sec).

Ohm (Ω) and is the unit of electrical resistance (volts/ampere).

Watts (W) and **kilowatts (kW)** are units of electric power.

Kilovolt-amperes (kVA) is a measurement of apparent electric power.

Kilowatt hour (kWhr) is a unit of electrical energy or work performed.

Joule (J) metric unit of energy = watt per sec.

$$1 \text{ Kwhr} = 2{,}655{,}000 \text{ ft-lb} = 1.341 \text{ hp-hr} = 3{,}413 \text{ Btu}$$
$$= 3{,}6000{,}000 \text{ joules}$$

Ohms Law Relationships (direct current)

$$E = IR = W/I = \sqrt{WR} \qquad W = I^2R = E^2/R = EI$$

$$I = E/R = W/E = \sqrt{W/R} \qquad R = E/I = W/I^2 = E^2/W$$

Electrical Formulas Symbols as above; plus

Eff = efficiency (expressed as a decimal)

pf = power factor (expressed as a decimal)

hp = horsepower output

Required	Direct Current	Alternating current Single-phase	Three-phase*
kVA		$\frac{IE}{1000}$	$\frac{1.73IE}{1000}$
Kilowatts	$\frac{IE}{1000}$	$\frac{IE(\mathrm{pf})}{1000}$	$\frac{1.73IE(\mathrm{pf})}{1000}$
Horsepower (output)	$\frac{IE(\mathrm{eff})}{1000}$	$\frac{IE(\mathrm{eff})(\mathrm{pf})}{746}$	$\frac{1.73IE(\mathrm{eff})(\mathrm{pf})}{746}$
Joules	$\frac{IE}{\mathrm{sec}}$	$\frac{IE(\mathrm{eff})(\mathrm{pf})}{\mathrm{sec}}$	$\frac{1.73IE(\mathrm{eff})(\mathrm{pf})}{\mathrm{sec}}$
Amperes (hp known)	$\frac{746\ (\mathrm{hp})}{E(\mathrm{eff})}$	$\frac{746\ (\mathrm{hp})}{E(\mathrm{eff})(\mathrm{pf})}$	$\frac{746\ (\mathrm{hp})}{1.73E(\mathrm{eff})(\mathrm{pf})}$
Amperes (kW known)	$\frac{1000\ \mathrm{kW}}{E}$	$\frac{1000\ \mathrm{kW}}{E(\mathrm{pf})}$	$\frac{1000\ \mathrm{kW}}{1.73E(\mathrm{pf})}$
Amperes (kVA known)		$\frac{1000\ \mathrm{kVA}}{E}$	$\frac{1000\ \mathrm{kVA}}{1.73E}$

*For three-phase systems E is measured line to line and I is phase current.

Miscellaneous Data

1 kilowatt-hour (kWhr) = 3,413 British thermal units (Btu)
1 Kilowatt (kW) = 1,000 watts (W)
1 Kilowatt (kW) = 1.341 horsepower (hp)
1 horsepower-hour (hp-hr) = 2,545 Btu
1 horsepower (hp) = 0.746 kilowatt (kW) or 746 watts (W)

MISCELLANEOUS ELECTRICAL FORMULAS

Ohms law:

Ohms = volts/amperes
Amperes = volts/ohms
Volts = amperes × ohms

Power—ac circuits:

$$\text{Efficiency} = \frac{746 \times \text{output horsepower}}{\text{input watts}}$$

Three-phase kilowatts =

$$\frac{\text{volts} \times \text{amperes} \times \text{power factor} \times 1.732}{1000}$$

Three-phase volt-amperes = volts × amperes × 1.732

Three-phase amperes =

$$\frac{746 \times \text{horsepower}}{1.732 \times \text{volts} \times \text{efficiency} \times \text{power factor}}$$

Three-phase efficiency =

$$\frac{746 \times \text{horsepower}}{\text{volts} \times \text{amperes} \times \text{power factor} \times 1.732}$$

$$\text{Three-phase power factor} = \frac{\text{input watts}}{\text{volts} \times \text{amperes} \times 1.732}$$

$$\text{Single-phase kilowatts} = \frac{\text{volts} \times \text{amperes} \times \text{power factor}}{1000}$$

$$\text{Single-phase amperes} = \frac{746 \times \text{horsepower}}{\text{volts} \times \text{efficiency} \times \text{power factor}}$$

$$\text{Single-phase efficiency} = \frac{746 \times \text{horsepower}}{\text{volts} \times \text{amperes} \times \text{power factor}}$$

$$\text{Single-phase power factor} = \frac{\text{input watts}}{\text{volts} \times \text{amperes}}$$

Horsepower (3-phase) =

$$\frac{\text{volts} \times \text{amperes} \times 1.732 \times \text{efficiency} \times \text{power factor}}{746}$$

Horsepower (1-phase) =

$$\frac{\text{volts} \times \text{amperes} \times \text{efficiency} \times \text{power factor}}{746}$$

Motor application formulas:

$$\text{Torque (lb-ft)} = \frac{\text{horsepower} \times 5250}{\text{rpm}}$$

$$\text{Horsepower} = \frac{\text{torque (lb-ft)} \times \text{rpm}}{5250}$$

Time for motor to reach operating speed (seconds):

$$\text{Seconds} = \frac{WK^2 \times \text{speed change}}{308 \times \text{avg. accelerating torque}}$$

WK^2 = inertia of rotor + inertia of load $(\text{lb-ft})^2$

$$\text{Average accelerating torque} = \frac{\frac{(\text{FLT} + \text{BDT}) + \text{BDT} + \text{LRT}}{2}}{3}$$

FLT = full load torque, BDT = breakdown torque,

LRT = locked rotor torque

$$\text{Load } WK^2 \text{ (at motor shaft)} = \frac{WK^2 \text{ (load)} \times \text{load rpm}^2}{\text{motor rpm}^2}$$

$$\text{Shaft stress (psi)} = \frac{\text{hp} \times 321{,}000}{\text{rpm} \times \text{shaft diam}^3}$$

For pumps:

$$\text{Horsepower} = \frac{\text{gpm} \times \text{head in feet} \times \text{specific gravity}}{3{,}960 \times \text{efficiency of pump}}$$

Power—dc circuits:

$$\text{Watts} = \text{volts} \times \text{amperes}$$

$$\text{Amperes} = \frac{\text{watts}}{\text{volts}}$$

$$\text{Horsepower} = \frac{\text{volts} \times \text{amperes} \times \text{efficiency}}{746}$$

For fans and blowers:

$$\text{Horsepower} = \frac{\text{CFM} \times \text{pressure (lb/ft}^2)}{33{,}000 \times \text{efficiency}}$$

Speed:

$$\text{Synchronous rpm} = \frac{\text{hertz} \times 120}{\text{Poles}}$$

$$\text{Percent slip} = \frac{\text{synchronous rpm} - \text{full load rpm} \times 100}{\text{synchronous rpm}}$$

Appendix

4

Mechanical Engineering Formula and Data

1. $W_a = \text{CFM} \times 60 \times W$

   $\text{Weight}_{(\text{pounds/hour})} = \text{cubic feet/min} \times 60\ \text{min/hr} \times \text{specific density}_{(\text{pound/cubic ft})}$

2. $W_a = \dfrac{\text{CFM} \times 60}{V_s}$

   $\text{Weight}_{(\text{pounds/hour})} = \dfrac{\text{cubic feet/min} \times 60\ \text{min/hr}}{\text{specific volume}_{(\text{cubic ft/lb})}}$

3. $Q_a = W_a \times (T_i - T_o) \times U_a$

   $\text{Quantity heat}_{(\text{Btu/hr})} = \text{weight}_{(\text{pounds/hr})} \times \text{temp diff.} \times \text{specific heat [air specific heat} = .24\ \text{Btu/lb-°F]}$

4. Heat transfer:

   $$\frac{I}{U} = \frac{I}{F_o} + \frac{L_1}{K_1} + \frac{L_2}{K_2} + \frac{I}{F_i}$$

U = heat transmittance, Btu/hr-ft^2-°F
F_o = outside air film coefficient
F_i = inside air film coefficient
L = thickness of material
K = conductivity of material (Btu/hr)(ft^2)(°F/T)

(Be sure that K and L are in the same thickness units; that is, in inches or feet.)

5. $Q = U(T_o - T_i)$

Q = heat flow (Btu/hr)(ft^2)
T_o = outside temperature
T_i = inside temperature
U = heat transmittance

6. Fan laws
where:

CFM = cubic feet per minute
CFM_1 = initial air flow rate
CFM_2 = final air flow rate
rpm = revolutions per minute
rpm_1 = initial fan speed
rpm_2 = final fan speed
SP = static pressure
SP_1 = initial system static pressure
SP_2 = final system static pressure
BHP = brake horsepower
BHP_1 = initial fan horsepower
BHP_2 = final fan horsepower
Fan size = constant
sp gr (air) = specific gravity of air is 1.0
Fan eff = fan efficiency = 65–85%
dia = diameter

$$\frac{CFM_2}{CFM_1} = \frac{rmp_2}{rmp_1} = \frac{dia_2}{dia_1}$$

$$\frac{SP_2}{SP_1} = \left(\frac{CFM_2}{CFM_1}\right)^2 = \left(\frac{rpm_2}{rpm_1}\right)^2$$

$$\frac{BPH_2}{BHP_1} = \left(\frac{CFM_2}{CFM_1}\right)^3 = \left(\frac{rpm_2}{rpm_1}\right)^3$$

$$BHP = \frac{CFM \times SP \times sp\ gr}{6{,}356 \times fan\ eff}$$

7. Pump laws
   where:

gpm = gallons per minute
  gpm_1 = initial gallon per minute
  GPM_2 = final gallon per minute
rpm = revolutions per minute
  rpm_1 = initial revolutions per minute
  rpm_2 = final revolutions per minute
HD = head pressure
BHP = brake horsepower
sp gr (water) = 1.0
Pump eff = 60–80%

$$\frac{gpm_2}{gpm_1} = \frac{rmp_2}{rmp_1}$$

$$\frac{HD_2}{HD_1} = \left(\frac{gpm_2}{gpm_1}\right)^2 = \left(\frac{rpm_2}{rpm_1}\right)^2$$

$$BHP = \frac{gpm \times HD \times sp\ gr}{3{,}960 \times pump\ eff}$$

8. Heat measures

1 calorie = the amount of heat needed to raise the temperature of one gram of water, one degree centigrade.
252 calories = 1 Btu
1 Btu = 778 ft-lb
1 Btu/lb = .55 cal/kg
1 Btu/lb = 2.326 kJ/kg
1 Btu/hr = 0.2931 watt

9. Water measures

1 U.S. gallon of water = 231 cubic inches (in^3)
1 U.S. gallon of water = 8.33 lb
1 U.S. gallon of water = 0.13368 cubic feet (ft^3)

1 cubic foot of water = 7.48 gal
1 cubic foot of water = 62.5 lb
1 foot of water = 0.434 psi

10. Water pressure
    1 cubic foot of water = 0.434 psi
    2.31 feet of water = 1 psi

11. Horsepower measures
    1 horsepower (hp) = 33,000 ft-lb/min
    1 horsepower = 746 watts
    1 horsepower = 778 ft-lb
    1 horsepower = 2,546 Btu/hr
    1 boiler (hp) = 33,475/hr
    1 boiler (hp) = 34.5 lb of steam/hr at 212°F

12. ASTM Btu values of fuels

Type	Fuel	Btu/Unit	Unit	Weight
Solids	Coal			
	Anthracite	11,000–13,000	pound	
	Bituminous	9,600–14,000	pound	
	Lignite	7,000	pound	
	Trash	7,500	pound	
	Wood	6,000	pound	
Liquids	Oil			
	#1	138,000	gallon	7.206 lb/gal
	#2	141,000	gallon	
	#3	144,000	gallon	
	#4	146,000	gallon	7.727 lb/gal
	#5	148,000	gallon	7.935 lb/gal
	#6	150,000	gallon	8.212 lb/gal
Gases				
	Natural gas	950–1150	cubic foot	
	Propane	91,500	gallon	

Appendix 5

Rules of Thumb

A. *COOLING LOAD RULES OF THUMB*

1. Offices, commercial	300–400 sq ft/ton:	30–40 Btu-hr/sq ft
large perimeter	225–275 sq ft/ton:	43–53 Btu-hr/sq ft
large, interior	300–350 sq ft/ton:	34–40 Btu-hr/sq ft
small	325–375 sq ft/ton:	32–37 Btu-hr/sq ft
2. Bank (main areas)	200–250 sq ft/ton:	48–60 Btu-hr/sq ft
3. Precision manufacturing	250–300 sq ft/ton:	40–48 Btu-hr/sq ft
4. Computer rooms	50–150 sq ft/ton:	80–240 Btu-hr/sq ft
5. Restaurants	100–250 sq ft/ton:	48–120 Btu-hr/sq ft
6. Cocktail lounges, bars	150–200 sq ft/ton:	60–80 Btu-hr/sq ft
7. Hospital patient rooms	250–300 sq ft/ton:	40–48 Btu-hr/sq ft
8. Buildings with 100% outside air systems (i.e., laboratories, hospitals)	100–300 sq ft/ton:	40–120 Btu-hr/sq ft
9. Medical centers	250–300 sq ft/ton:	40–48 Btu-hr/sq ft
10. Residential	500–700 sq ft/ton:	17–24 Btu-hr/sq ft
11. Apartments (Eff., 1 room, 2 room)	350–450 sq ft/ton:	27–34 Btu-hr/sq ft
12. Hotel guest rooms	250–300 sq ft/ton:	40–48 Btu-hr/sq ft
13. Motels, dormitories	400–500 sq ft/ton:	24–30 Btu-hr/sq ft
14. School classrooms	225–275 sq ft/ton:	43–53 Btu-hr/sq ft
15. Dining halls, lunch rooms	100–250 sq ft/ton:	48–120 Btu-hr/sq ft
16. Libraries, museums	250–350 sq ft/ton:	34–48 Btu-hr/sq ft
17. Retail, department stores	200–250 sq ft/ton:	48–60 Btu-hr/sq ft
18. Specialty shops, drug stores	175–225 sq ft/ton:	53–69 Btu-hr/sq ft
19. Super markets	250–350 sq ft/ton:	34–48 Btu-hr/sq ft

20. Malls	150–350 sq ft/ton: 34–80 Btu-hr/sq ft
21. Auditoriums, theaters	0.05–0.07 tons/seat
22. Churches	0.04–0.06 tons/seat
23. Bowling alleys	1.5–2.5 tons/alley
24. Perimeter spaces	1.0–3.0 CFM/sq ft
25. Interior spaces	0.5–1.5 CFM/sq ft
26. Building block CFM	1.0–1.5 CFM/sq ft

Cooling load calculations should be conducted using industry accepted methods to determine actual cooling load requirements.

B. *HEATING LOAD RULES OF THUMB*

1. All buildings and spaces	20–60 Btu-hr/sq ft 25–40 Btu-hr/sq ft average
2. Buildings with 100% Outside Air Systems (i.e., laboratories, hospitals)	40–120 Btu-hr/sq ft
3. Bldg w/heavy insul., few windows	AC Tons × 12,000 Btu-hr/ton × 1.2
4. Bldg w/light insul., many windows	AC Tons × 12,000 Btu-hr/ton × 1.5
5. Walls below grade	−30°F–6.0 Btu-hr/sq ft −10°F–4.0 Btu-hr/sq ft 0°F–3.0 Btu-hr/sq ft 10°F–2.0 Btu-hr/sq ft 30°F–1.5 Btu-hr/sq ft
6. Floors below grade	−30°F–3.0 Btu-hr/sq ft −10°F–2.0 Btu-hr/sq ft 0°F–1.5 Btu-hr/sq ft 10°F–1.0 Btu-hr/sq ft 30°F–0.5 Btu-hr/sq ft

Heating load calculations should be conducted using industry accepted methods to determine actual heating load requirements.

C. *INFILTRATION RULES OF THUMB*

1. Heating infiltration (15 mph wind)

a. Air change (AC) rate method	Range, 0–10 AC/hr
for commercial buildings	1.0 AC/hr, 1 exterior wall 1.5 AC/hr, 2 exterior walls 2.0 AC/hr, 3 or 4 exterior walls
for vestibules	3.0 AC/hr
b. CFM/sq ft of wall method	Range, 0 1.0 CFM/sq ft
• Tight buildings	0.1 CFM/sq ft
• Average buildings	0.3 CFM/sq ft
• Leaky buildings	0.6 CFM/sq ft

c. Crack method
Range, 0.12–2.8 CFM/ft of crack
Average, 1.0 CFM/ft of crack

2. Cooling infiltration (7.5 mph wind):
Cooling load infiltration is generally ignored unless close tolerances in temperature and humidity control are required. Cooling infiltration values are generally taken as ½ of the values listed above for heating infiltration.
3. No infiltration losses or gains for rooms below grade or for interior spaces.
4. Buildings which are not humidified have no latent infiltration heating load.
5. Winter sensible infiltration loads will generally be ½ to 3 times the conduction heat losses (average 1.0–2.0 times).

D. *VENTILATION RULES OF THUMB*

1. Outdoor air
a. American Society of Heating, Refrigerating, and Air Conditioning Engineers Standard 62
Most common range 15–35 CFM/person minimum
Average 15 CFM/person
Smoking lounges 60 CFM/person minimum
b. Building Officials and Code Administrators Code 5 CFM/person minimum
c. Southern Building Code Congress International Code 5 CFM/person minimum
d. Uniform Building Code 5 CFM/person minimum
e. Indoor Air Quality (IAQ)—ASHRAE Standard 62
1) Causes of poor IAQ:
Inadequate ventilation problems 50% of all IAQ problems are due to lack of ventilation, such as:
Intermittent airflow
Poor air distribution
Poor intake/exhaust locations
Inadequate operation
Inadequate maintenance
2) IAQ control methods:
Control temperature and humidity
Ventilation—dilution
Remove pollution source
Filtration
3) IAQ factors:
Thermal environment
Smoke
Odors
Irritants—dust

Stress problems (perceptible, non-perceptible)
Toxic gases—carbon monoxide, carbon dioxide
Allergens—pollen
Biological contaminants—bacteria, old, pathogens, *Legionella,* microorganisms, fungi

2. Toilet rooms 2.0 CFM/sq ft
   10 (air changes)/hr
   100 CFM/water closet and urinal
   a. American Society of Heating, Refrigerating, and Air-Conditioning Engineers Standard 62 — 50 CFM/water closet and urinal
   b. BOCA Code — 75 CFM/water closet and urinal
   c. SBCCI Code — 2.0 CFM/sq ft
   d. UBC Code — 5.0 AC/hr
3. Mechanical rooms 2 CFM/sq ft
   10 CFM/BHP; 1 BHP = 34,500 Btu-hr
   (8 CFM/BHP combustion air, 2 CFM/BHP ventilation—Cleaver Brooks)
4. Electrical rooms 2.0 CFM/sq ft
   10 Air change/hr
   5 CFM/kVA of transformer
5. Combustion air sources
   a. BOCA Code
      1) Inside air 1 sq in/1000 Btu-hr
      2) Outside air 1 sq in/4000 Btu-hr w/o horz. ducts
         1 sq in/2000 Btu-hr w/horz. ducts
   b. SBCCI Code
      1) Solid fuels 2 sq in/1000 Btu-hr; 200 sq in minimum
      2) Liquid and gas fuels
      3) Confined spaces
         a. Inside air 1 sq in/1000 Btu-hr; 100 sq in minimum
         b. Outside air 1 sq in/4000 Btu-hr w/o horz. ducts
            1 sq in/2000 Btu-hr w/horz. ducts
      4) Unconfined spaces
         Outside air 1 sq in/5000 Btu-hr
   c. UBC Code
      1) Confined spaces
         a. Inside air 1 sq in/1000 Btu-hr; each opening
         b. Outside air 1 sq in/4000 Btu-hr w/o horz. ducts
            1 sq in/2000 Btu-hr w/horz. ducts
      2) Unconfined spaces
         Outside air 1 sq in/5000 Btu-hr
   d. National Fire Protection Association 54—National Fuel Gas Code
      1) Confined spaces

a. Inside air — 1 sq in/1000 Btu-hr; 100 sq in min.
b. Outside air — 1 sq in/4000 Btu-hr; direct communication with outside
1 sq in/4000 Btu-hr w/vert. ducts
1 sq in/2000 Btu-hr w/horz. ducts

2) Unconfined spaces
a. Tight buildings — As specified for confined spaces.
b. Leaky buildings — Infiltration may be adequate.

E. *HUMIDIFICATION RULES OF THUMB*

1. Single-pane windows — 10% RH maximum
Double-pane windows — 30% RH maximum
Triple-pane windows — 40% RH maximum
Proper vapor barriers and moisture control must be provided to prevent moisture condensation in walls and to prevent mold, fungi, bacteria, and other plant and microorganism growth (based on 0°F outside design temperature).
2. Human comfort — 30–60% RH (relative humidity)
Electrical equipment, computers — 35–55% RH
3. Winter design relative humidities
Outdoor air below 32°F — 70–80% RH
Outdoor air 32–60°F — 50% RH

F. *PEOPLE / OCCUPANCY RULES OF THUMB*

1. Offices, commercial	
General	80–150 sq ft/person
Private	1, 2, or 3 people
Private	100–150 sq ft/person
Conference, meeting rooms	20–50 sq ft/person
2. Bank (main areas)	5–150 sq ft/person
3. Precision manufacturing	100–300 sq ft/person
4. Computer rooms	80–150 sq ft/person
5. Restaurants	15–50 sq ft/person
6. Cocktail lounges, bars	15–50 sq ft/person
7. Hospital patient rooms	25–75 sq ft/person
8. Hospital general areas	50–150 sq ft/person
9. Medical centers	50–100 sq ft/person
10. Residential	200–600 sq ft/person
11. Apartments (Eff., 1 room, 2 room)	100–400 sq ft/person
12. Hotel guest rooms	100–200 sq ft/person
13. Motels, dormitories	100–200 sq ft/person
14. School classrooms	20–30 sq ft/person

15. Dining halls, lunch rooms — 10–50 sq ft/person
16. Libraries, museums — 30–100 sq ft/person
17. Retail, department stores — 15–75 sq ft/person
18. Specialty shops, drug stores — 15–50 sq ft/person
19. Super markets — 50–100 sq ft/person
20. Malls — 50–100 sq ft/person
21. Auditoriums, theaters — 5–20 sq ft/person
22. Churches — 5–20 sq ft/person
23. Bowling alleys — 2–6 People/lane

People/occupancy requirements should be determined from the architect or client whenever possible.

G. *LIGHTING RULES OF THUMB*

1. Offices, commercial	
General	1.5–3.0 watts/sq ft
Private	2.0–5.0 watts/sq ft
Conference, meeting rooms	2–6.0 watts/ sq ft
2. Bank (main areas)	2.0–5.0 watts/sq ft
3. Precision manufacturing	3.0–10.0 watts/sq ft
4. Computer rooms	1.5–5.0 watts/sq ft
5. Restaurants	1.5–3.0 watts/sq ft
6. Cocktail lounges, bars	1.5–2.0 watts/sq ft
7. Hospital patient rooms	1.5–2.5 watts/sq ft
8. Hospital general areas	1.5–2.5 watts/sq ft
9. Medical centers	1.5–2.5 watts/sq ft
10. Residential	1.0–4.0 watts/sq ft
11. Apartments (Eff., 1 room, 2 room)	1.0–4.0 watts/sq ft
12. Hotel guest rooms	1.0–3.0 watts/sq ft
13. Motels, dormitories	1.0–3.0 watts/sq ft
14. School classrooms	2.0–6.0 watts/sq ft
15. Dining halls, lunch rooms	1.5–2.5 watts/sq ft
16. Libraries, museums	1.0–3.0 watts/sq ft
17. Retail, department stores	2.0–6.0 watts/sq ft
18. Specialty shops, drug stores	1.0–3.0 watts/sq ft
19. Super markets	1.0–3.0 watts/sq ft
20. Malls	1.0–2.5 watts/sq ft
21. Auditoriums, theaters	1.0–3.0 watts/sq ft*
22. Churches	1.0–3.0 watts/sq ft
23. Bowling alleys	1.0–2.5 watts/sq ft

*Does not include theatrical lighting.

Actual lighting layouts should be used for calculating lighting loads whenever available.

H. *APPLIANCE / EQUIPMENT RULES OF THUMB*

1. Offices, commercial — 0.5–2.0 watts/sq ft

2. Computer rooms 2.0–50.0 watts/sq ft

Actual equipment layouts should be used for calculating equipment loads; for example:

Movie projectors, slide projectors, overhead projectors, etc., can generally be ignored because lights are off when being used and lighting load will normally be larger than this equipment.

Items such as coffeepots, microwave ovens, refrigerators, food warmers, etc., should be considered when calculating equipment loads.

Kitchen, laboratory, hospital, and computer room equipment should be obtained from the owner or architect due to extreme variability of equipment loads.

I. *COOLING LOAD FACTORS*

1. Diversity factors—the engineer's judgment is applied to various people, lighting, equipment, and total loads when considering actual usage. Actual diversities may vary depending on building type and occupancy. Diversities listed here are for office buildings and similar facilities.
   a. Room/spaces peak loads

People	1.0 × calc. load
Lights	1.0 × calc. load
Equipment	1.0 × calc. load*

   *Calculated (calc.) load may have the diversity factor calculated with individual pieces of equipment or as a group or not at all.

   b. Floor/zone block loads

People	0.90 × sum of peak room/spaces people loads
Lights	0.95 × sum of peak room/spaces lighting loads
Equipment	0.90 × sum of peak room/spaces equipment
Loads floor/zone total loads	0.90 × sum of peak room/spaces total loads

   c. Building block loads

People	0.75 × sum of peak room/spaces people loads
Lights	0.95 × sum of peak room/spaces lighting loads
Equipment	0.75 × sum of peak room/spaces equipment loads
Building total loads	0.85 × sum of peak room/spaces total loads

2. Safety factors

a. Room/space peak loads	1.1 × calc. load
b. Floor/zone loads (sum of peak)	1.0 × calc. load
c. Floor/zone loads (block)	1.1 × calc. load
d. Building loads (sum of peak)	1.0 × calc. load
e. Building loads (block)	1.1 × calc. load

3. Cooling load factors

a. Lighting load factors
   Fluorescent lights 1.25 × bulb watts
   Incandescent lights 1.00 × bulb watts
   HID lighting 1.25 × bulb watts
b. Return air plenum (RAP) factors
   Heat of lights to space with RAP 0.76 × lighting load
   Heat of lights to RAP 0.24 × lighting load
   Heat of roof to space with RAP 0.30 × roof load
   Heat of roof to RAP 0.70 × roof load
c. Ducted exhaust or return air (DERA) factors
   Heat of lights to space with DERA 1.0 × lighting load
   Heat of roof to space with DERA 1.0 × roof load
d. Other cooling load factors (CLF) are in accordance with ASHRAE recommendations.
   CLF × other loads

J. *HEATING FACTORS*

1. Safety factors
   a. Room/space peak loads 1.1 × calc. load
   b. Floor/zone loads (sum of peak) 1.0 × calc. load
   c. Floor/zone loads (block) 1.1 × calc. load
   d. Building loads (sum of peak) 1.0 × calc. load
   e. Building loads (block) 1.1 × calc. load
   f. Generally sum of peak loads = 1.1 × block loads
2. Heating load credits
   a. Solar—credit for solar gains should not be taken unless building is specifically designed for solar heating. Solar gain is not a factor at night when design temperatures generally reach their lowest point.
   b. People—credit for people should not be taken. People gain is not a factor at night when design temperatures generally reach their lowest point, because buildings are usually unoccupied at night.
   c. Lighting—credit for lighting should not be taken. Lighting is an inefficient means to heat a building, and lights are usually off at night when design temperatures generally reach their lowest point.
   d. Equipment—credit for equipment should not be taken unless a reliable source of heat is generated 24 hours a day (e.g., computer facility, industrial process). Only a portion of this load should be considered (50%), and the building heating system should be able to keep the building from freezing if these equipment loads are shut down for extended periods of time. Consider what would happen if the system or process would shut down for extended periods of time.

Appendix

6

Measurement Data

MEASUREMENTS

LENGTH

12 inches = 1 foot
3 feet = 1 yard
5,280 feet = 1 statute mile
6,080 feet = 1 nautical mile
1 mil = 0.001 inch

AREA

144 square inches = 1 square foot
9 square feet = 1 square yard
Cross-sectional area in circular mils = square of diameter in mils

VOLUME

1,728 cubic inches = 1 cubic foot
27 cubic feet = 1 cubic yard

WEIGHT

16 ounces = 1 pound
2,000 pounds = 1 short ton
2,240 pounds = 1 long ton

CIRCULAR MEASURE

60 seconds = 1 minute
60 minutes = 1 degree
360 degrees = 1 circle
90 degrees = 1 right angle
11¼ degrees = 1 point on the compass

TIME

60 seconds = 1 minute
60 minutes = 1 hour
24 hours = 1 day
365 days = 1 year

CONVERSION FACTORS

Atmosphere (standard) = 29.92 inches of mercury
Atmosphere (standard) = 14.7 pounds per square inch
1 horsepower = 746 watts
1 horsepower = 33,000 foot-pounds of work per minute
1 British thermal unit (Btu) = 778 foot-pounds
1 cubic foot = 7.48 gallons
1 gallon = 231 cubic inches
1 cubic foot of fresh water = 62.5 pounds
1 cubic foot of salt water = 64 pounds
1 foot of head pressure of water = 0.434 pounds per square inch
1 inch of head pressure of mercury = 0.491 pounds per square inch
1 gallon of fresh water = 8.33 pounds
1 barrel (oil) = 42 gallons
1 long ton of fresh water = 36 cubic feet
1 long ton of salt water = 25 cubic feet
1 ounce (avoirdupois) = 437.5 grains

THERM-HOUR CONVERSION FACTORS

1 therm-hour = 100,000 Btu per hour
1 brake horsepower = 2,544 Btu per hour

$$1 \text{ brake horsepower} = \frac{2544}{100,000} = 0.02544 \text{ therm-hour}$$

$$1 \text{ therm-hour} = \frac{100,000}{2,544} = 39.3082 \text{ brake horsepower}$$

(40 hp is close enough)

$$1 \text{ therm-hour} = \frac{100,000}{33,475} = 2.873 \text{ boiler horsepower}$$

(3 hp is close enough)

EXAMPLE: How many therm-hours are in a 100-hp engine?
ANSWER: $100 \times 0.02544 = 2.544$ therm-hours

BOILER HORSEPOWER

1 boiler hp = 33,475 Btu per hour
= 34.5 lb steam per hour at 212°F
= 139 sq ft EDR (equivalent direct radiation)
1 EDR = 240 Btu per hour
1 kW = 3,413 Btu per hour

WEIGHT OF WATER AT 62°F

1 cu in = 0.0361 lb (of water)
1 cu ft = 62.355 lb
1 gal = 8.3391 lb ($8\frac{1}{3}$ is close enough)

MISCELLANEOUS DATA

3,413 British thermal units (Btu) = 1 kilowatt-hour (kWhr)
1,000 watts = 1 kilowatt (kW)
1.341 horsepower = 1 kilowatt
2.545 Btu = 1 horsepower-hour (hp-hr)
0.746 kilowatt = 1 hp
1 micron (μm) = one millionth of a meter (unit of length)

Appendix

7

Conversion Data

To Convert	into	Multiply by
	A	
acres	sq feet	43,560.0
acres	sq meters	4,047.0
acres	sq miles	1.562×10^{-3}
acres	sq yards	4,840
acre-feet	cu feet	43,560.0
acre-feet	gallons	3.259×10^{5}
amperes/sq cm	amps/sq in	6.452
amperes/sq cm	amps/sq meter	10^{4}
amperes/sq in	amps/sq cm	0.1550
amperes/sq in	amps/sq meter	1,550.0
amperes/sq meter	amps/sq cm	10^{-4}
amperes/sq meter	amps/sq in	6.452×10^{-4}
amperes-hours	coulombs	3,600.0
amperes-hours	faradays	0.03731
amperes-turns	gilberts	1.257
amperes-turns/cm	amp-turns/in	2.540
amperes-turns/cm	amp-turns/meters	100.0
amperes-turns/cm	gilberts/cm	1.257
amperes-turns/in	amp-turns/cm	0.3937
amperes-turns/in	amp-turns/meters	39.37
amperes-turns/in	gilberts/cm	0.4950
amperes-turns/meter	amp-turns/cm	0.01
amperes-turns/meter	amp-turns/in	0.0254

To Convert	into	Multiply by
amperes-turns/meter	gilberts/cm	0.01257
acres	sq meters	100.0
atmospheres	cm of mercury	76.0
atmospheres	ft of water (at 4°C)	33.90
atmospheres	in of mercury (at °C)	29.92
atmospheres	kg/sq cm	1.0333
atmospheres	kg/sq meter	10,332
atmospheres	pounds/sq in	14.70
atmospheres	tons/sq ft	1.058

B

To Convert	into	Multiply by
barrels (oil)	gallons (oil)	42.0
bars	atmospheres	0.9869
bars	dynes/sq cm	10^6
bars	kg/sq meter	1.020×10^4
bars	pounds/sq ft	2,089.0
bars	pounds/sq in	14.50
Btu	ergs	1.0550×10^{10}
Btu	foot-lbs	778.3
Btu	gram-calories	252.0
Btu	kilocalories	0.252
Btu	horsepower-hrs	3.931×10^{-4}
Btu	joules	1,054.8
Btu	kilogram-calories	0.2520
Btu	kilogram-meters	107.5
Btu	kilowatt-hr	2.928×10^{-4}
Btu/hr	foot-pounds/sec	0.2162
Btu/hr	gram-cal/sec	0.0700
Btu/hr	horsepower-hrs	3.929×10^{-4}
Btu/hr	watts	0.2931
Btu/min	foot-lbs/sec	12.96
Btu/min	horsepower	0.02356
Btu/min	kilowatts	0.01757
Btu/min	watts	17.57
Btu/sq ft/min	watts/sq in	0.1221
bushels	cu ft	1.2445
bushels	cu in	2,150.4
bushels	cu meters	0.03524
bushels	liters	35.24
bushels	pecks	4.0
bushels	pints (dry)	64.0
bushels	quarts (dry)	32.0

To Convert	into	Multiply by
	C	
centares (centiares)	sq meters	1.0
centigrade	Fahrenheit	$(C^\circ \times 9/5) + 32$
centigrams	grams	0.01
centiliters	liters	0.01
centimeters	feet	3.281×10^{-2}
centimeters	inches	0.3937
centimeters	kilometers	10^{-5}
centimeters	meters	0.01
centimeters	miles	6.214×10^{-4}
centimeters	millimeters	10.0
centimeters	mils	393.7
centimeters	yards	1.094×10^{-2}
centimeters-dynes	cm-grams	1.020×10^{-3}
centimeters-dynes	pound-feet	7.376×10^{-8}
centimeters-grams	cm-dynes	980.7
centimeters-grams	meter-kg	10^{-5}
centimeters-grams	pound-feet	7.233×10^{-5}
centimeters of mercury	atmospheres	0.01316
centimeters of mercury	feet of water	0.4461
centimeters of mercury	kg/sq meter	136.0
centimeters of mercury	pounds/sq ft	27.85
centimeters of mercury	pounds/sq in	0.1934
centimeters/sec	feet/min	1.9685
centimeters/sec	feet/sec	0.03281
centimeters/sec	kilometers/hr	0.036
centimeters/sec	knots	0.0194
centimeters/sec	meters/min	0.6
centimeters/sec	miles/hr	0.02237
centimeters/sec	miles/min	3.728×10^{-4}
centimeters/sec/sec	feet/sec/sec	0.03281
centimeters/sec/sec	km/hr/sec	0.036
centimeters/sec/sec	meters/sec/sec	0.01
centimeters/sec/sec	miles/hr/sec	0.02237
circular mils	sq cm	5.067×10^{-6}
circular mils	sq mils	0.7854
circular mils	sq inches	7.854×10^{-7}
coulombs	faradays	1.036×10^{-5}
coulombs/sq cm	coulombs/sq in	64.52
coulombs/sq cm	coulombs/sq meter	10^{4}
coulombs/sq in	coulombs/sq cm	0.1550
coulombs/sq in	coulombs/sq meters	1,550.0
coulombs/sq meter	coulombs/sq cm	10^{-4}

To Convert	into	Multiply by
coulombs/sq meter	coulombs/sq in	$6,452 \times 10^{-4}$
cubic centimeters	cu feet	3.531×10^{-5}
cubic centimeters	cu inches	0.06102
cubic centimeters	cu meters	10^{-6}
cubic centimeters	cu yards	1.308×10^{-6}
cubic centimeters	gallons (U.S. liq.)	$2,642 \times 10^{-4}$
cubic centimeters	liters	0.001
cubic centimeters	pints (U.S. liq.)	2.113×10^{-3}
cubic centimeters	quarts (U.S. liq.)	1.057×10^{-3}
cubic feet	bushels (dry)	0.08036
cubic feet	cu centimeters	28.320.0
cubic feet	cu inches	1,728.0
cubic feet	cu meters	0.02832
cubic feet	cu yards	0.03704
cubic feet	gallons (U.S. liq.)	7.48052
cubic feet	liters	28.32
cubic feet	pints (U.S. liq.)	59.84
cubic feet	quarts (U.S. liq.)	29.92
cubic feet/min	cu cm/sec	472.0
cubic feet/min	gallons/sec	0.1247
cubic feet/min	liters/sec	0.4720
cubic feet/min	pounds of water	62.43
cubic feet/sec	millions gals/day	0.646317
cubic feet/sec	gallons/min	148.81
cubic inches	cu cm	16.0
cubic inches	cu feet	5.787×10^{-4}
cubic inches	cu meters	1.639×10^{-5}
cubic inches	cu yards	2.143×10^{-5}
cubic inches	gallons (U.S. liq.)	4.329×10^{-3}
cubic inches	liters	0.01639
cubic inches	mil-feet	1.061×10^{5}
cubic inches	pints (U.S. liq.)	0.03463
cubic inches	quarts (U.S. liq.)	0.01732
cubic meters	bushels (dry)	28.38
cubic meters	cu cm	10^{6}
cubic meters	cu feet	35.31
cubic meters	cu yards	1.308
cubic meters	gallons (U.S. liq.)	264.2
cubic meters	liters	1,000.0
cubic meters	pints (U.S. liq.)	2,113.0
cubic meters	quarts (U.S. liq.)	1,057.0
cubic yards	cu cm	7.646×10^{5}
cubic yards	cu feet	27.0
cubic yards	cu inches	46,656.0
cubic yards	cu meters	0.7646

To Convert	into	Multiply by
cubic yards	gallons (U.S. liq.)	202.0
cubic yards	liters	764.6
cubic yards	pints (U.S. liq.)	1,615.9
cubic yards	quarts (U.S. liq.)	807.9
cubic yards/min	cubic ft/sec	0.45
cubic yards/min	gallons/sec	3.367
cubic yards/min	liters/sec	12.74
	D	
days	hours	24.0
days	minutes	1,440.0
days	seconds	86,400.0
decigrams	grams	0.1
deciliters	liters	0,1
decimeters	meters	0.1
degrees (angle)	minutes	60.0
degrees (angle)	quandrants	0.01111
degrees (angle)	radians	0.01745
degrees (angle)	seconds	3,600.0
degrees/sec	radians/sec	0.01745
degrees/sec	revolutions/min	0.1667
degree/sec	revolutions/sec	2.778×10^{-3}
dekagrams	grams	10.0
dekaliters	liters	10.0
dekameters	meters	10.0
drams	grams	1.7718
drams	grains	27.3437
drams	ounces	0.0625
dynes	grams	1.020×10^{-3}
dynes	joules/cm	10^{-7}
dynes	joules/meter (newtons)	10^{-5}
dynes	kilograms	1.020×10^{-6}
dynes	poundals	7.2333×10^{-5}
dynes	pounds	2.248^{-6}
dynes/sq cm	Bars	10^{-6}
	E	
ergs	Btu	9.480×10^{-11}
ergs	dynes-centimeters	1.0
ergs	foot-pounds	7.367×10^{-8}
ergs	gram-calories	0.2389×10^{-7}
ergs	grams-cm	1.020×10^{-3}
ergs	horsepower-hrs	3.7250×10^{-14}

To Convert	into	Multiply by
ergs	joules	10^{-7}
ergs	kg-calories	2.389×10^{-11}
ergs	kg-meters	1.020×10^{-8}
ergs	kilowatt-hrs	0.2778×10^{-13}
ergs	watt-hours	0.2778×10^{-10}
ergs/sec	Btu/min	$5{,}688 \times 10^{-9}$
ergs/sec	ft-lbs/min	4.427×10^{-6}
ergs/sec	ft-lbs/sec	7.3756×10^{-8}
ergs/sec	horsepower	1.341×10^{-10}
ergs/sec	kg-calories/min	1.433×10^{-9}
ergs/sec	kilowatts	10^{-10}

F

To Convert	into	Multiply by
farads	microfarads	10^{6}
faradays	ampere-hours	26.80
faradays	coulombs	9.649×10^{-4}
fathoms	feet	6.0
feet	centimeters	30.48
feet	kilometers	3.048×10^{-4}
feet	meters	0.3048
feet	miles (naut.)	1.645×10^{-4}
feet	miles (stat.)	1.894×10^{-4}
feet	millimeters	304.8
feet	mils	1.2×10^{4}
feet of water	atmospheres	0.02950
feet of water	in of mercury	0.8826
feet of water	kg/sq cm	0.03048
feet of water	pounds/sq ft	62.43
feet of water	pounds/sq in	0.4335
feet/min	cm/sec	0.5080
feet/min	feet/sec	0.01667
feet/min	kms/hr	0.01829
feet/min	meters/min	0.3048
feet/min	miles/hr	0.01136
feet/sec	cm/sec	30.48
feet/sec	kms/hr	1.097
feet/sec	knots	0.5921
feet/sec	meters/min	18.29
feet/sec	miles/hr	0.6818
feet/sec	miles/min	0.01136
feet/sec/sec	cm/sec/sec	30.48
feet/sec/sec	kms/hr/sec	1.097
feet/sec/sec	meters/sec/sec	0.3048
feet/sec/sec	miles/hr/sec	0.6818

To Convert	into	Multiply by
feet/100 feet	per cent grade	1.0
foot-pounds	Btu	1.286×10^{-3}
foot-pounds	ergs	1.356×10^{7}
foot-pounds	gram-calories	0.3238
foot-pounds	hp-hrs	5.050×10^{-7}
foot-pounds	joules	1.356
foot-pounds	kg-calories	3.24×10^{-4}
foot-pounds	kg-meters	0.1383
foot-pounds	kilowatt-hrs	3.766×10^{-7}
foot-pounds/min	Btu/min	1.286×10^{-3}
foot-pounds/min	foot-pounds/sec	0.01667
foot-pounds/min	horsepower	3.030×10^{-5}
foot-pounds/min	kg-calories/min	3.24×10^{-4}
foot-pounds/min	kilowatts	2.260×10^{-5}
foot-pounds/sec	Btu/hr	4.6263
foot-pounds/sec	Btu/min	0.07717
foot-pounds/sec	horsepower	1.818×10^{-3}
foot-pounds/sec	kg-calories/min	0.01945
foot-pounds/sec	kilowatts	1.356×10^{-3}
furlongs	rods	40.0
furlongs	feet	660.0

G

To Convert	into	Multiply by
gallons	cu cm	3,785.0
gallons	cu feet	0.1337
gallons	cu inches	231.0
gallons	cu meters	3.785×10^{-3}
gallons	cu yards	4.951×10^{-3}
gallons	liters	3.785
gallons	pints	8.0
gallons	quarts	4.0
gallons (liq. Br. Imp.)	gallons (U.S. liq.)	1.20095
gallons (U.S.)	gallons (Imp.)	0.83267
gallons of water	pounds of water	8.3453
gallons/min	cu ft/sec	2.228×10^{-3}
gallons/min	liters/sec	0.06308
gallons/min	cu ft/hr	8.0208
gausses	lines/sq in	6.452
gausses	webers/sq cm	10^{-8}
gausses	webers/sq in	6.452×10^{-8}
gausses	webers/sq meter	10^{-4}
gilberts	ampere-turns	0.7958
gilberts/cm	amp-turns/cm	0.7958
gilberts/cm	amp-turn/in	2.021

To Convert	into	Multiply by
gilberts/cm	amp-turns/meter	79.58
gills	liters	0.1183
gills	pints (liq.)	0.25
grains (troy)	grains (avdp)	1.0
grains (troy)	grams	0.06480
grains (troy)	ounces (avdp)	2.0833×10^{-3}
grains (troy)	pennyweight (troy)	0.04167
grains/U.S. gal	parts/million	17.118
grains/U.S. gal	pounds/million gal	142.86
grains/Imp. gal	parts/million	14.286
grams	dynes	980.7
grams	grains	15.43
grams	joules/cm	9.807×10^{-5}
grams	joules/meter (newtons)	980.7×10^{-3}
grams	kilograms	0.001
grams	milligrams	1,000
grams	ounces (avdp)	0.03527
grams	ounces (troy)	0.03215
grams	poundals	0.07093
grams	pounds	2.205×10^{-3}
grams/cm	pounds/inch	$5{,}600 \times 10^{-3}$
grams/cu cm	pounds/cu ft	62.43
grams/cu cm	pounds/cu in	0.03613
grams/cu cm	pounds/mil-foot	3.405×10^{-7}
grams/liter	grains/gal	58.417
grams/liter	pounds/1,000 gal	8.3445
grams/liter	pounds cu ft	0.062427
grams/liter	parts/millon	1,000.0
gram-calories	Btu	3.968×10^{-3}
gram-calories	foot-pounds	3.08
gram-calories	horsepower-hrs	1.5586×10^{-6}
gram-calories	kilowatt-hrs	1.1×10^{-6}
gram-calories	watt-hrs	1.1310×10^{-3}
gram-calories	Btu/hr	14.28
gram-centimeters	Btu	9.2×10^{-3}
gram-centimeters	ergs	980.7
gram-centimeters	joules	9.80×10^{-5}
gram-centimeters	kg-cal	$2.3\ 10^{-3}$
gram-centimeters	kg-meters	10^{-5}

H

To Convert	into	Multiply by
hectares	acres	2.47
hectares	sq feet	1.076×10^{5}

To Convert	into	Multiply by
hectograms	grams	100.0
hectoliters	liters	100.0
hectometers	meters	100.0
hectowatts	watts	100.0
henries	millihenries	1,000.0
horsepower	Btu/min	42.44
horsepower	foot-lbs/min	33,000.0
horsepower	foot-lbs/sec	550.0
horsepower (metric) (542.5 ft lb/sec)	horsepower (550 ft lb/sec)	0.9863
horsepower (550 ft lb/sec	horsepower (metric) (542.5 ft lb/sec)	1.014
horsepower	kg-calories/min	10.68
horsepower	kilowatts	0.7457
horsepower	watts	745.7
horsepower (boiler)	Btu/hr	33.479
horsepower	kilowatts	9.803
horsepower-hrs	Btu	2,547
horsepower-hrs	ergs	2.6845×10^{13}
horsepower-hrs	foot-lbs	1.98×10^{6}
horsepower-hrs	gram-calories	641,190.0
horsepower-hrs	joules	2.684×10^{6}
horsepower-hrs	kg-calories	641.1
horsepower-hrs	kg-meters	2.737×10^{5}
horsepower-hrs	kilowatt-hrs	0.7457
hours	days	4.167×10^{-2}
hours	minutes	60.0
hours	seconds	3,600.0
hours	week	5.952×10^{-3}
	I	
inches	centimeters	2.540
inches	feet	8.333×10^{-2}
inches	meters	2.540×10^{-2}
inches	miles	1.578×10^{-5}
inches	millimeters	25.40
inches	mils	1,000.0
inches	yards	2.778×10^{-2}
inches of mercury	atmospheres	3.342×10^{-2}
inches of mercury	feet of water	1.133
inches of mercury	kg/sq cm	3.453×10^{-2}
inches of mercury	kg/sq meter	345.3
inches of mercury	pounds/sq ft	70.73

To Convert	into	Multiply by
inches of mercury	pounds/sq in	0.4912
inches of water (at 4°C)	atmospheres	2.458×10^{-3}
inches of water (at 4°C)	inches of mercury	7.355×10^{-3}
inches of water (at 4°C)	kg/sq cm	2.450×10^{-3}
inches of water (at 4°C)	ounces/sq in	0.5781
inches of water (at 4°C)	pounds/sq ft	5.204
inches of water (at 4°C)	pounds/sq in	3.613×10^{-2}
	J	
joules	Btu	9.480×10^{-4}
joules	ergs	10^{7}
joules	foot-pounds	0.7376
joules	kg-calories	2.389×10^{-4}
joules	kg-meters	0.01020
joules	watt-hrs	2.778×10^{-4}
joules/cm	grams	1.020×10^{4}
joules/cm	dynes	10^{7}
joules/cm	joules/meter (newtons)	100.0
joules/cm	poundals	723.3
joules/cm	pounds	22.48
	K	
kilograms	dynes	980,665.0
kilograms	grams	1,000.0
kilograms	joules/cm	0.09807
kilograms	joules/meter (newton)	9.807
kilograms	poundals	70.93
kilograms	pounds	2.205
kilograms	tons (long)	$9{,}842 \times 10^{-4}$
kilograms	tons (short)	1.102×10^{-3}
kilograms/cu meter	grams/cu cm	0.001
kilograms/cu meter	pounds/cu ft	0.06243
kilograms/cu meter	pounds cu in	3.613×10^{-5}
kilograms/cu meter	pounds/mil-foot	3.405×10^{-10}
kilograms/meter	pounds/ft	0.6720
kilograms/sq cm	atmospheres	0.9678
kilograms/sq cm	feet of water	32.81
kilograms/sq cm	inches of mercury	28.96
kilograms/sq cm	pounds/sq ft	2,048
kilograms/sq cm	pounds/sq in	14.22
kilograms/sq meter	atmospheres	9.678×10^{-5}
kilograms/sq meter	bars	98.07×10^{-6}

To Convert	into	Multiply by
kilograms/sq meter	feet of water	3.281×10^{-3}
kilograms/sq meter	inches of mercury	2.896×10^{-3}
kilograms/sq meter	pounds/sq ft	0.2048
kilograms/sq meter	pounds/sq in	1.422×10^{-3}
kilograms/sw mm	kg/sq meter	10^{6}
kilograms-calories	Btu	3.968
kilograms-calories	foot-pounds	3.088.0
kilograms-calories	hp-hrs	1.560×10^{-3}
kilograms-calories	joules	4,186.0
kilograms-calories	kg-meters	426.9
kilograms-calories	kilojoules	4.186
kilograms-calories	kilowatt-hrs	1.163×10^{-3}
kilograms meters	Btu	9.294×10^{-3}
kilograms meters	ergs	9.804×10^{7}
kilograms meters	foot-pounds	7.233
kilograms meters	kg-calories	2.342×10^{-3}
kilograms meters	kilowatt-hrs	2.723×10^{-6}
kilolines	maxwells	1,000.0
kiloliters	liters	1,000.0
kilometers	centimeters	10^{5}
kilometers	feet	3,281.0
kilometers	inches	3.937×10^{4}
kilometers	meters	1000.0
kilometers	miles	0.6214
kilometers	millimeters	10^{6}
kilometers	yards	1,094
kilometers/hr	cm/sec	27.78
kilometers/hr	feet/min	54.68
kilometers/hr	feet/sec	0.9113
kilometers/hr	knots	0.5396
kilometers/hr	meters/min	16.67
kilometers/hr	miles/hr	0.6214
kilometers/hr/sec	cm/sec/sec	27.78
kilometers/hr/sec	ft/sec/sec	0.9113
kilometers/hr/sec	meters/sec/sec	0.2778
kilometers/hr/sec	miles/hr/sec	0.6214
kilowatts	Btu/min	56.92
kilowatts	foot-lbs/sec	737.6
kilowatts	horsepower	1.341
kilowatts	kg-calories/min	14.34
kilowatts	watts	1,000.0
kilowatts-hrs	Btu	3,413.0
kilowatts-hrs	ergs	$3{,}600 \times 10^{13}$
kilowatts-hrs	foot-lbs	2.655×10^{6}

To Convert	into	Multiply by
kilowatts-hrs	Btu	3,413.0
kilowatts-hrs	ergs	$3{,}600 \times 10^{13}$
kilowatts-hrs	foot-lbs	2.655×10^{6}
kilowatts-hrs	gram-calories	859,850.0
kilowatts-hrs	Btu	3,413.0
kilowatts-hrs	horsepower-hrs	1,341.0
kilowatts-hrs	joules	3.6×10^{6}
kilowatts-hrs	kg-calories	859.85
kilowatts-hrs	kg-meters	3.671×10^{5}
kilowatts-hrs	pounds of water evaporated from and at 212°F	3.53
kilowatts-hrs	pounds of water raised from 62° to 212°F	22.75
knots	feet/hr	6,080.0
knots	kilometers/hr	1.8532
knots	nautical miles/hr	1.0
knots	statute miles/hr	1.151
knots	yards/hr	2,027.0
knots	feet/sec	1.689

L

To Convert	into	Multiply by
league	miles (approx.)	3.0
lines/sq cm	gausses	1.0
lines/sq in	gausses	0.1550
lines/sq in	webers/sq cm	1.550×10^{-9}
lines/sq in	webers sq in	10^{-6}
lines/sq in	webers/sq meter	1.550×10^{-5}
links (engineer's)	inches	12.0
links (surveyor's)	inches	7.92
liters	bushels (U.S. dry)	0.02838
liters	cu cm	1,000.0
liters	cu feet	0.03531
liters	cu inches	61.02
liters	cu meters	0.001
liters	cu yards	1.308×10^{-3}
liters	gallons (U.S. liq.)	0.2642
liters	pints (U.S. liq.)	2.113
liters	quarts (U.S. liq.)	1.057
liters/min	cu ft/sec	5.886×10^{-4}
liters/min	gals/sec	4.403×10^{-3}
lumens/sq ft	foot-candles	1.0
lux	foot-candles	0.0929

To Convert	into	Multiply by
	M	
maxwells	kilolines	0.001
maxwells	webers	10^{-8}
megalines	maxwells	10^{6}
megohms	microhms	10^{12}
megohms	ohms	10^{6}
meters	centimeters	100.0
meters	feet	3.281
meters	inches	39.37
meters	kilometers	0.001
meters	miles (naut.)	5.396×10^{-4}
meters	miles (stat.)	6.214×10^{-4}
meters	millimeters	1,000.0
meters	yards	1.094
meters	varas	1.179
meters/min	cm/sec	1.667
meters/min	feet/min	3.281
meters/min	feet/sec	0.05468
meters/min	kms/hr	0.06
meters/min	miles/hr	0.03728
meters/sec	feet/sec	3.281
meters/sec	kilometers/hr	3.6
meters/sec	kilometers/min	0.06
meters/sec	miles/hr	2.237
meters/sec	miles/min	0.03728
meters/sec/sec	cm/sec/sec	100.0
meters/sec/sec	ft/sec/sec	3.281
meters/sec/sec	kms/hr/sec	3.6
meters/sec/sec	miles/hr/sec	2.237
meter-kilograms	cm-dynes	9.807×10^{7}
meter-kilograms	cm-grams	10^{5}
meter-kilograms	pound-feet	7.233
microfarad	farads	10^{-6}
micrograms	grams	10^{-6}
microhms	megohms	10^{-12}
microhms	ohms	10^{-6}
microliters	liters	10^{-6}
miles (naut.)	feet	6,076.103
miles (naut.)	kilometers	1.852
miles (naut.)	meters	1,852.0
miles (naut.)	miles (statute)	1.1508
miles (naut.)	yards	2.025.4
miles (statute)	centimeters	1.609×10^{5}
miles (statute)	feet	5,280.0

To Convert	into	Multiply by
miles (statute)	inches	$6,336 \times 10^{4}$
miles (statute)	kilometers	1.609
miles (statute)	meters	1,609.0
miles (statute)	miles (naut.)	0.86879
miles (statute)	yards	1,760.0
miles/hr.	cm/sec	44.70
miles/hr.	feet/min	88.0
miles/hr.	feet/sec	1.467
miles/hr.	kms/hr	1.609
miles/hr.	kms/min	0.02682
miles/hr.	knots	0.08684
miles/hr.	meters/min	26.82
miles/hr.	miles/min	0.01667
miles/hr/sec	cm/sec/sec	44.70
miles/hr/sec	feet/sec/sec	1.467
miles/hr/sec	kms/hr/sec	1.609
miles/hr/sec	meters/sec/sec	0.4470
miles/min	cm/sec	2,682.0
miles/min	feet/sec	88.0
miles/min	kms/min	1.609
miles/min	miles (naut.)/min	0.8684
miles/min	miles/hr	60.0
mil-feet	cu inches	9.425×10^{-6}
milliers	kilograms	1,000
milligrams	grams	0.001
milligrams/liter	parts/million	1.0
millihenries	henries	0.001
milliters	liters	0.001
millimeters	centimeters	0.1
millimeters	feet	3.281×10^{-3}
millimeters	inches	0.03937
millimeters	kilometers	10^{-6}
millimeters	meters	0.001
millimeters	miles	6.21×10^{-7}
millimeters	mils	39.37
millimeters	yards	1.094×10^{-3}
million gals/day	cu ft/sec	1.54723
mils	centimeters	2.540×10^{-3}
mils	feet	8.333×10^{-5}
mils	inches	0.001
mils	kilometers	2.540×10^{-8}
mils	yards	2.778×10^{-5}
miner's inches	cu ft/min	1.5
minutes (angles)	degrees	0.01667
minutes (angles)	quandrants	1.852×10^{-4}

To Convert	into	Multiply by
minutes (angles)	radians	2.909×10^{-4}
minutes (angles)	seconds	60.0
myriagrams	kilograms	10.0
myriagrams	kilograms	10.0
myriawatts	kilowatts	10.0
	N	
nepers	decibels	8.686
	O	
ohms	megohms	10^{-6}
ohms	microhms	10^{6}
ounces	drams	16.0
ounces	grains	437.5
ounces	grams	28.349527
ounces	pounds	0.0625
ounces	ounces (troy)	0.9115
ounces	tons (long)	2.790×10^{-5}
ounces	tons (metric)	2.835×10^{-5}
ounces (fluid)	cu inches	1.805
ounces (fluid)	liters	0.02957
ounces (troy)	grains	480.0
ounces (troy)	grams	31.103481
ounces (troy)	ounces (avdp)	1.09714
ounces (troy)	pennyweights (troy)	20.0
ounces (troy)	pounds (troy)	0.08333
ounces/sq in	pounds/sq in	0.0625
	P	
parts/million	grains/U.S. gal	0.0584
parts/million	grains/Imp. gal	0.07016
parts/million	pounds/million gal	8.345
pennyweights (troy)	grains	24.0
pennyweights (troy)	ounces (troy)	0.05
pennyweights (troy)	grams	1.55517
pennyweights (troy)	pounds (troy)	4.1667×10^{-3}
pints (dry)	cu inches	33.60
pints (liq.)	cu cm	473.2
pints (liq.)	cu feet	0.01671
pints (liq.)	cu inches	28.87
pints (liq.)	cu meters	4.732×10^{-4}
pints (liq.)	cu yards	6.189×10^{-4}

To Convert	into	Multiply by
pints (liq.)	gallons	0.125
pints (liq.)	liters	0.4732
pints (liq.)	quarst (liq.)	0.5
poundals	dynes	13,826.0
poundals	grams	14.10
poundals	joules/cm	1.383×10^{-3}
poundals	joules/meter (newtons)	0.1383
poundals	kilograms	0.01410
poundals	pounds	0.03108
pounds	drams	256.0
pounds	dynes	44.4823×10^{4}
pounds	grains	7,000.0
pound	grams	453.5924
pounds	joules/cm	0.04448
pounds	joules/meter (newton)	4.448
pounds	kilograms	0.4536
pounds	ounces	16.0
pounds	ounces (troy)	14.5833
pounds	poundals	32.17
pounds	pounds (troy)	1.21528
pounds	tons (short)	0.0005
pounds (troy)	grains	5,760.0
pounds (troy)	grams	373.24177
pounds (troy)	ounces (avdp)	13.1657
pounds (troy)	ounces (troy)	12.0
pounds (troy)	pennyweights (troy)	240.0
pounds (troy)	pounds (avdp)	0.822587
pounds (troy)	tons (long)	3.6735×10^{-4}
pounds (troy	tons (metric)	3.7324×10^{-4}
pounds (troy)	tons (short)	4.1143×10^{-4}
pounds of water	cu feet	0.01602
pounds of water	cu inches	27.68
pounds of water	gallons	0.1198
pounds of water/min	cu ft/sec	2.670×10^{-4}
pound-feet	cm-dynes	1.356×10^{7}
pound-feet	cm-grams	13,825.0
pound-feet	meter-kg	0.1383
pounds/cu ft	grams/cu cm	0.01602
pounds/cu ft	kg/cu meter	16.02
pounds/cu ft	pounds/cu in	5.787×10^{-4}
pounds/cu ft	pounds/mil-foot	5.456×10^{-9}
pounds/cu in	gms/cu cm	27.68
pounds/cu in	kg/cu meter	2.768×10^{4}
pounds/cu in	pounds/cu ft	1,728.0
pounds/cu in	pounds/mil-foot	9.425×10^{-6}

To Convert	into	Multiply by
pounds/ft	kg/meter	1.488
pounds/in	gms/cm	178.6
pounds/mil-foot	gms/cu cm	2.306×10^6
pounds/sq ft	atmospheres	4.725×10^{-4}
pounds/sq ft	feet of water	0.01602
pounds/sq ft	inches of mercury	0.01414
pounds/sq ft	kg/sq meter	4.882
pounds/sq ft	pounds/sq in	6.944×10^{-3}
pounds/sq in	atmospheres	0.6804
pounds/sq in	feet of water	2.307
pounds/sq in	kg/sq meter	703.1
pounds/sq in	pounds/sq ft	144.0
	Q	
quadrants (angle)	degrees	90.0
quadrants (angle)	minutes	5,400.0
quadrants (angle)	radians	1.571
quadrants (angle)	seconds	3.24×10^5
quarts (dry)	cu inches	67.20
quarts (liq.)	cu cm	946.4
quarts (liq.)	cu feet	0.03342
quarts (liq.)	cu inches	57.75
quarts (liq.)	cu meters	9.464×10^{-4}
quarts (liq.)	cu yards	1.238×10^{-3}
quarts (liq.)	gallons	0.25
quarts (liq.)	liters	0.9463
	R	
radians	degrees	57.30
radians	minutes	3,438.0
radians	quadrants	0.6366
radians	seconds	2.063×10^5
radians/sec	degrees/sec	57.30
radians/sec	revolutions/min	9.549
radians/sec	revolutions/sec	0.1592
radians/sec/sec	rev/min/min	573.0
radians/sec/sec	rev/min/sec	9.549
radians/sec/sec	revs/sec/sec	0.1592
revolutions	degrees	360.0
revolutions	quadrants	4.0
revolutions	radians	6.283
revolutions/min	degrees/sec	6.0
revolutions/min	radians/sec	0.1047

To Convert	into	Multiply by
revolutions/min	revs/sec	0.01667
revolutions/min/min	radians/sec/sec	1.745×10^{-3}
revolutions/min/min	revs/min/sec	0.01667
revolutions/min/min	revs/sec/sec	2.778×10^{-4}
revolutions/sec	degrees/sec	360.0
revolutions/sec	radians/sec	6.283
revolutions/sec	revs/min	60.0
revolutions/sec/sec	radians/sec/sec	6.283
revolutions/sec/sec	revs/min/min	3,600.0
revolutions/sec/sec	revs/min/sec	60.0
rods	feet	16.5
	S	
seconds (angle)	degrees	2.778×10^{-4}
seconds (angle)	minutes	0.01667
seconds (angle)	quandrants	3.087×10^{-6}
seconds (angle)	radians	4.848×10^{-6}
square centimeters	circular mils	1.973×10^{5}
square centimeters	sq feet	1.076×10^{-3}
square centimeters	sq inches	0.1550
square centimeters	sq meters	0.0001
square centimeters	sq miles	3.861×10^{-11}
square centimeters	sq millimeters	100.0
square centimeters	sq yards	1.196×10^{-4}
square feet	acres	2.296×10^{-5}
square feet	circular mils	1.833×10^{8}
square feet	sq cm	929.0
square feet	sq inches	144.0
square feet	sq meters	0.09290
square feet	sq miles	3.587×10^{-8}
square feet	sq millimeters	9.290×10^{4}
square feet	sq yards	0.1111
square inches	circular mils	1.273×10^{6}
square inches	sq cm	6.452
square inches	sq feet	6.944×10^{-3}
square inches	sq millimeters	645.2
square inches	sq mils	10^{6}
square inches	sq yards	7.716×10^{-4}
square kilometer	acres	247.1
square kilometers	sq cm	10^{10}
square kilometers	sq ft	10.76×10^{6}
square kilometers	sq inches	1.550×10^{9}
square kilometers	sq meters	10^{6}

To Convert	into	Multiply by
square kilometers	sq miles	0.3861
square kilometers	sq yards	1.196×10^{6}
square meters	acres	2.471×10^{-4}
square meters	sq cm	10^{4}
square meters	sq feet	10.76
square meters	sq inches	1,550.0
square meters	sq miles	3.861×10^{-7}
square meters	sq millimeters	10^{6}
square meters	sq yards	1.196
square miles	acres	640.0
square miles	sq feet	27.88×10^{6}
square miles	sq kms	2.590
square miles	sq meters	2.590×10^{6}
square miles	sq yards	3.098×10^{6}
square millimeters	circular mils	1,973.0
square millimeters	sq cm	0.01
square millimeters	sq feet	1.076×10^{-5}
square millimeters	sq inches	1.550×10^{-3}
square mils	circular mils	1.273
square mils	sq cm	6.452×10^{-6}
square mils	sq inches	10^{-6}
square yards	acres	2.066×10^{-4}
square yards	sq cm	8,361.0
square yards	sq feet	9.0
square yards	sq inches	1,296
square yards	sq meters	0.8361
square yards	sq miles	3.228×10^{-7}
square yards	sq millimeters	8.361×10^{5}

T

To Convert	into	Multiply by
temperature (°C) +273	absolute temperature (°C)	1.0
temperature (°C) +17.78	temperture (°F)	1.8
temperature (°F) +460	absolute temperature (°F)	1.0
temperature (°F)-32	temperature (°C)	5/9
tons (long)	kilograms	1,016
tons (long)	pounds	2,240
tons (long)	tons (short)	1.120
tons (metric)	kilograms	1,000
tons (metric)	pounds	2,205.0
tons (short)	kilograms	907.1848

To Convert	into	Multiply by
tons (short)	ounces	32,000.0
tons (short)	ounces (troy)	29,166.66
tons (short)	pounds	2,000
tons (short)	pounds (troy)	2,430.56
tons (short)	tons (long)	0.892897
tons (short)	tons (metric)	0.9078
tons (short)/sq ft	kg/sq meter	9,765.0
tons (short)/sq ft	pounds/sq in	2,000
tons of water/24 hrs	pounds of water/hr	83.333
tons of water/24hrs	gallons/min	0.16643
	W	
watts	Btu/hr	3.413
watts	Btu/min	0.05688
watts	ergs/sec	10^7
watts	foot-lbs/min	44.27
watts	foot-lbs/sec	0.7378
watts	horsepower	1.341×10^{-3}
watts	horsepower (metric)	1.360×10^{-3}
watts	kg-calories/min	0.01433
watts	kilowatts	0.001
watt-hours	Btu	3.413
watt-hours	ergs	3.60×10^{10}
watt-hours	foot-pounds	2,656.0
watt-hours	gram-calories	859.85
watt-hours	horsepower-hrs	1.341×10^{-3}
watt-hours	kilogram-calories	0.8598
watt-hours	kilogram-meters	367.2
watt-hours	kilowatt-hrs	0.001
webers	maxwells	10^8
webers	lilolines	10^5
webers/sq in	gausses	1.550×10^7
webers/sq in	webers/sq cm	0.1550
webers/sq in	weber/sq meter	1,550.0
webers/sq meter	gausses	10^4
webers/sq meter	lines/sq in	6.452×10^4
webers/sq meter	webers/sq cm	10^{-4}
webers/sq meter	webers/sq in	6.452×10^{-4}
	Y	
yards	centimeters	91.44
yards	feet	3.0
yards	inches	36.0

To Convert	into	Multiply by
yards	kilometers	9.144×10^{-4}
yards	meters	0.9144
yards	miles (naut.)	4.937×10^{-4}
yards	miles (stat.)	5.682×10^{-4}
yards	millimeters	914.4

ABOUT THE AUTHORS

BERNARD T. LEWIS is a recognized leader in facilities management and the author of McGraw-Hill's *Facility Manager's Operation and Maintenance Handbook* and *Facility Manager's Portable Handbook.* Dr. Lewis owns and manages a facilities management consultant firm that specializes in facilities engineering management problem-solving. His clients have included the Charles E. Smith Companies, Georgetown University, George Washington University, and the Port Authority of New York. A member of the Association of Facilities Engineers, Dr. Lewis teaches engineering management, lectures on facilities management nationally and internationally, and is the author of 21 books and numerous magazine articles on facilities management and engineering. A West Point graduate in civil/mechanical engineering, he holds an M.A. in mathematics from Columbia University and a Ph.D. from Pacific Western University.

RICHARD P. PAYANT has more than 7 years of experience as Director of Facilities Management at Georgetown University and more than 23 years' experience with the Army Corps of Engineers. A certified plant engineer and facility manager, an adjunct professor at George Mason University, and the author of several professional publications, he has an M.A. in business management/public administration from Central Michigan University and a B.S. from Norwich University.

Index

Made in the USA
Lexington, KY
24 March 2017